SpringerBriefs in Architectural Design and Technology

Series Editor

Indexed by SCOPUS

Understanding the complex relationship between design and technology is increasingly critical to the field of Architecture. The *Springer Briefs in Architectural Design and Technology* series provides accessible and comprehensive guides for all aspects of current architectural design relating to advances in technology including material science, material technology, structure and form, environmental strategies, building performance and energy, computer simulation and modeling, digital fabrication, and advanced building processes. The series features leading international experts from academia and practice who provide in-depth knowledge on all aspects of integrating architectural design with technical and environmental building solutions towards the challenges of a better world. Provocative and inspirational, each volume in the Series aims to stimulate theoretical and creative advances and question the outcome of technical innovations as well as the far-reaching social, cultural, and environmental challenges that present themselves to architectural design today. Each brief asks why things are as they are, traces the latest trends and provides penetrating, insightful and in-depth views of current topics of architectural design. *Springer Briefs in Architectural Design and Technology* provides must-have, cutting-edge content that becomes an essential reference for academics, practitioners, and students of Architecture worldwide.

Simona Mannucci

Climate Adaptation in Urban Planning

Toward Sustainable and Resilient Urban Environments

Springer

Simona Mannucci
Department of Civil, Constructional
and Environmental Engineering (DICEA)
Sapienza Università di Roma
Roma, Italy

ISSN 2199-580X ISSN 2199-5818 (electronic)
SpringerBriefs in Architectural Design and Technology
ISBN 978-981-97-4105-2 ISBN 978-981-97-4106-9 (eBook)
https://doi.org/10.1007/978-981-97-4106-9

This Springer imprint is published by the registered company Springer Nature Singapore Pte Ltd.
The registered company address is: 152 Beach Road, #21-01/04 Gateway East, Singapore 189721, Singapore

A Matilde

Acknowledgments

I sincerely thank Prof. Carlo Cecere, Prof. Marco Ferrero, Prof. Jan H. Kwakkel, Prof. Federica Rosso, and Prof. Michele Morganti for their support.

Special appreciation is also due to Prof. Jan H. Kwakkel and Dr. Bramka A. Jafino for their invaluable assistance and work with the case study presented in Chap. 4. Their expertise significantly enriched the depth and quality of the research.

On a personal note, I am very thankful to my family for their unwavering encouragement and support. Last but not least, I thank my husband for putting up with me for better or for worse; you are the true example of "resilience."

Contents

Chapter 1
Introduction

Abstract The impacts of climate change, such as global warming, rising sea levels, and increased extreme weather events, pose significant threats to urban areas. These challenges necessitate a reevaluation of urban design, construction, and management, highlighting the need for comprehensive climate action. Despite advancements, many cities struggle to meet climate targets. Traditional planning paradigms, based on historical assumptions of stationarity, are inadequate in addressing the dynamic and uncertain nature of climate change and urban delvelopment. In this book is discussed the evolution of planning paradigms from modernism to contemporary approaches that emphasize sustainability, resilience, adaptation, and regeneration, exploring the integration with innovative methods to manage complexity and uncertainty. The book aims to support planners in navigating these challenges, particularly in flood resilience, by exploring cross-disciplinary methods and tools for effective long-term planning.

Keywords Climate change · Resilience · Built environment · Climate adapation · Multi-disciplinary approaches

1.1 Introduction

The escalating impacts of climate change, characterized by global warming, rising sea levels, and ecological disruptions, are widely recognized threats to urbanized areas. Stemming from both natural processes and anthropogenic activities, climate change presents a multifaceted challenge with far-reaching consequences for our environment, economy, health, and security. Indeed, the increasing frequency of extreme weather events, ranging from floods to unprecedented heatwaves and droughts, poses significant challenges to the sustainability and resilience of urban environments.

These challenges necessitate a profound reevaluation of how cities are designed, constructed, and managed, highlighting the critical need for comprehensive action to mitigate and adapt to climate change. Despite advancements in addressing climate

S. Mannucci, *Climate Adaptation in Urban Planning*,
SpringerBriefs in Architectural Design and Technology,
https://doi.org/10.1007/978-981-97-4106-9_1

change, several European cities still face challenges in meeting the targets set forth in the Paris Agreement and the 2030 Agenda [1, 2].

In response, transformative initiatives like the European Green Deal [3], aiming for net-zero greenhouse gas emissions by 2050 and emphasizing a green and digital transition, and the New European Bauhaus initiative [4], which bridges science, technology, art, and culture to co-create solutions for a sustainable and inclusive future, have been introduced. These focus on sustainability, aesthetics, and inclusion. However, planning and designing cities, taking into account the built environment's inherent complexity and the unpredictable changes induced by climate change, represents a fundamental challenge for urban and architectural planners. This challenge is difficult to address with traditional planning paradigms and outdated tools.

Conventionally, planning has relied on the assumption of stationarity. However, anthropogenic and climatic changes have significantly altered the conditions underlying historical forecasting methods [5]. In planning terms, this means plans based on historical assumptions may not achieve their intended objectives if the future deviates from these assumptions. While overlooking uncertainties may simplify short-term planning, it increases the risk of long-term adverse surprises.

Since the latter half of the twentieth century, the principles of modernism in architecture and urban planning [6] have significantly influenced urban contexts worldwide. Despite evolving architectural and planning methodologies, the belief in the transformative power of architecture and planning has remained constant [7].

As climate change intensifies, urban planners and designers are at a pivotal juncture, requiring a departure from traditional paradigms to more dynamic, integrated approaches. This evolution underlines the necessity of blending existing models with innovative strategies to manage increasing complexity and uncertainty, ensuring that our urban planning and design frameworks are resilient and responsive to the changing climate for sustainable urban futures [8].

From this perspective, the book aims to support planners and designers in navigating the complexities of climate change adaptation within urban and architectural planning, focusing on flood resilience. It seeks to explore and synthesize cross-disciplinary methods and tools for effective long-term planning under uncertainty, moving from traditional paradigms to innovative, integrated approaches that address the multifaceted challenges of the built environment and climate change.

Specifically, the book is organized as follows:

First, Chap. 2 frames the evolution of planning paradigms for the built environment in the context of climate change. It traces the evolution of planning paradigms from sustainability to resilience, adaptation, and regeneration [9–12]. Moreover, the chapter emphasizes the importance of acknowledging the complexity of the built environment, a critical factor that persists beyond the specific terminology used to describe these planning approaches. Additionally, it explores the potential of innovative planning methodologies while addressing the practical challenges of applying these approaches in real-world urban settings.

Chapter 3 delves into the translation of planning paradigms into concrete design solutions, with a specific emphasis on addressing climate change through the lens of flood management [13]. It outlines how technical and technological solutions are formulated to mitigate the impacts of floods, examining strategies applied within the built environment across various scales—from national frameworks to individual building designs. The chapter also focuses on the gaps in attaining a comprehensive understanding of long-term adaptation strategies.

Chapter 4 explores the application of scenario planning and Decision Making under Deep Uncertainty (DMDU) to urban resilience and climate adaptation. It begins with an overview of scenario planning, including its advantages and limitations for addressing challenges in the built environment. The chapter proceeds to delve into DMDU, detailing a standard workflow for its application, which encompasses case study selection, data gathering, and scenario analysis. A particular case study focusing on a flood-prone area in Rome is discussed, showcasing the uncertainties associated with land use and infrastructure impacts. Furthermore, the chapter identifies the challenges in converting the insights gained into practical strategies and calls for further research to fully exploit these methodologies for developing adaptive and resilient urban planning approaches.

Finally, Chap. 5 concludes the discussion on the evolution of planning paradigms in architecture and urban planning, offering insights into future directions. It emphasizes the shift toward dynamic strategies that effectively address the challenges posed by climate change. Additionally, the chapter advocates for the enhancement of educational programs to better equip future planners with the skills necessary to navigate complex and uncertain conditions.

References

1. D. Reckien et al., How are cities planning to respond to climate change? Assessment of local climate plans from 885 cities in the EU-28. J. Clean. Prod. **191**, 207–219 (2018)
2. United Nations (UN), *Transforming Our World: The 2030 Agenda for Sustainable Development* (2015)
3. European Commission, *The European Green Deal* (Brussels, 2019)
4. European Commission, *Level(s) and the New European Bauhaus* (Publications Office of the European Union, 2022)
5. A.P.C.D. Milly et al., Stationarity is dead: stationarity whither water management? Science **319**(5863), 573–574 (2008)
6. E.P. Mumford, *The CIAM Discourse on Urbanism, 1928–1960* (MIT Press, Cambridge, 2002)
7. S. Richards, *Architect Knows Best: Environmental Determinism in Architecture Culture From 1956 to the Present* (Routledge, December 2014), pp. 1–178
8. W. Rauws, Embracing uncertainty without abandoning planning: exploring an adaptive planning approach for guiding urban transformations. DisP Plan Rev **53**(1), 32–45 (2017)
9. D. Michalina, P. Mederly, H. Diefenbacher, B. Held, Sustainable urban development: a review of urban sustainability indicator frameworks. Sustainability **13**(16), 1–20 (2021)
10. A. Wardekker, Contrasting the framing of urban climate resilience. Sustain. Cities Soc. **75**(July), 103258 (2021)
11. T.K. Dhar, L. Khirfan, Climate change adaptation in the urban planning and design research: missing links and research agenda. J. Environ. Plan. Manag. **60**(4), 602–627 (2017)
12. C. Du Plessis, Towards a regenerative paradigm for the built environment. Build. Res. Inf. **40**(1), 7–22 (2012)
13. M. Radhakrishnan, A. Pathirana, R. M. Ashley, B. Gersonius, C. Zevenbergen, Flexible adaptation planning for water sensitive cities. Cities **78**, 87–95 (2018)

Chapter 2
Planning Paradigms to Navigate Complexity and Climate Change

Abstract In response to the escalating challenges of climate change, urban planning has evolved from traditional models to embrace innovative, adaptive strategies. This chapter examines the evolution of planning paradigms, highlighting the shift towards climate-responsive planning and the recognition of the built environment as a complex system. It explores concepts such as urban resilience, adaptive planning, and urban regeneration, emphasizing their differences and intersections. The chapter aims to provide a comprehensive understanding of how acknowledging the complexity of the built environment has influenced new planning paradigms and addresses the challenges of implementing these paradigms in real-world urban contexts. It offers insights into current planning practices essential for fostering resilient and sustainable cities.

Keywords Climate change · Urban planning · Built environment · Sustainability urban resilience · Adaptive planning · Urban regeneration · Complex systems · Dynamic strategies · Climate-responsive planning

2.1 Climate-Responsive Urban Planning

In the face of escalating climate change challenges, planning for the built environment has transitioned from traditional models to a dynamic field that embraces innovative and adaptive strategies. This chapter delves into the evolution of planning paradigms in response to climate change, focusing on the shift from conventional methods to innovative, adaptive strategies for urban planning. It discusses the significance of climate-responsive planning, the pivotal role of sustainability, and the recognition of the built environment as a complex system. Additionally, the chapter explores concepts such as urban resilience, adaptive planning, and urban regeneration, highlighting the differences and intersections among these concepts. It aims to provide a comprehensive understanding of how the recognition of the built environment's complexity has influenced new planning paradigms and addresses the challenges of implementing these paradigms in real-world urban contexts.

S. Mannucci, *Climate Adaptation in Urban Planning*,
SpringerBriefs in Architectural Design and Technology,
https://doi.org/10.1007/978-981-97-4106-9_2

By synthesizing these diverse approaches, the chapter seeks to offer a comprehensive perspective on current planning practices for cities in the face of climate change. As urban environments evolve within a changing climate, understanding these contemporary practices becomes imperative for fostering resilient and sustainable cities worldwide.

2.2 The Turning Point of Sustainability

The term "sustainability" gained prominence in the late twentieth century in response to escalating environmental concerns, highlighting the need for a more balanced and enduring approach to development. The 1972 Stockholm Declaration marked the first significant step toward the modern understanding of sustainability by linking environmental objectives with economic development drivers. It underscored the interconnectedness of proposed solutions, including considerations for governance, technology transfer, and the recognition of cultural and economic differences among nations and stakeholders. Although this pioneering document offered a comprehensive and holistic view, recognizing sustainability's dynamic and systems-oriented nature, it somewhat underestimated the impact of technology and the interconnected effects of evolving culture, public policy, and economic markets [1].

The concept of "sustainable development" was officially introduced in 1987 with the publication of the Brundtland Report by the World Commission on Environment and Development [2], defining it as "*development that meets the needs of the present without compromising the ability of future generations to meet their own needs.*" The United Nations institutionalization of sustainable development through the Brundtland Report positioned economic growth as a solution to ecological and social challenges; this approach, which emphasizes win–win solutions, tends to depoliticize sustainability by treating economic, social, and environmental goals as equally essential. Echoing many aspects of the Stockholm Declaration, the Brundtland Report highlighted the critical challenges related to economic consumption, production, and development in achieving sustainability. It stressed the need to bridge cooperation gaps, enhance financial resources, and improve technology accessibility for a broader implementation of sustainable solutions globally. Furthermore, it outlined initial steps toward a large-scale transition to a low-carbon global economy, including insights on how industries could improve efficiency and *"produce more with less"* by developing and deploying new technologies.

Broadly speaking, the concept of sustainability is built on three principal pillars (Fig. 2.1): environmental, societal, and economic sustainability [3].

Environmental sustainability involves the responsible stewardship of natural resources and ecosystems [5]. It acknowledges the interconnectedness of human activities with the environment, recognizing that humans' well-being and ecosystems' health are intricately linked. Consequently, any disturbance to the balance

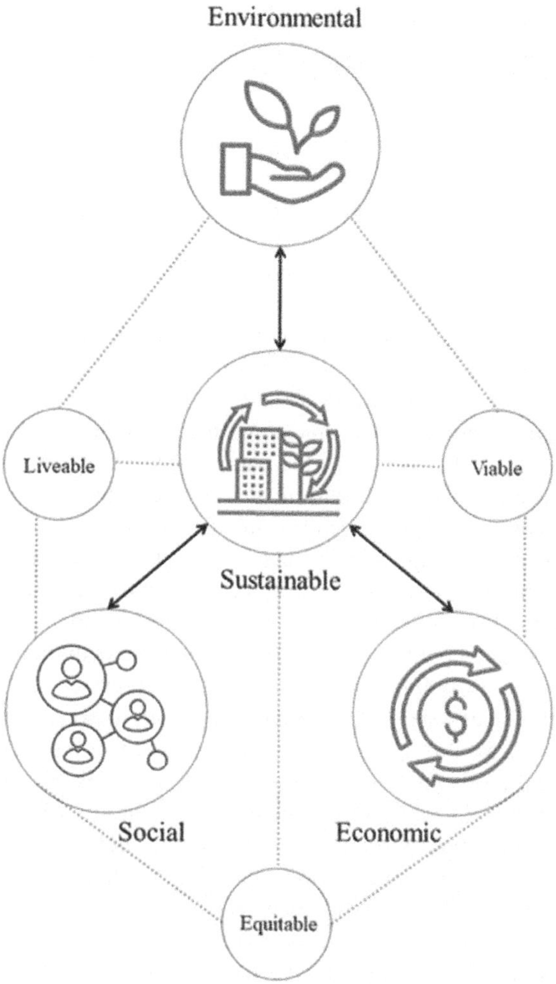

Fig. 2.1 Sustainability pillars. Author's adaptation from [4]

between the two can have far-reaching consequences. This interconnectedness underscores the importance of adopting holistic and systemic approaches to address environmental challenges [6].

Furthermore, environmental sustainability emphasizes the prudent use of resources, ensuring that extraction rates do not exceed the Earth's capacity for regeneration. Sustainable practices aim to minimize the depletion of non-renewable resources and promote the responsible management of renewable resources, fostering a harmonious coexistence between human activities and the planet's ecological systems. Integral components of environmental sustainability include pollution mitigation and waste reduction. Strategies in this area focus on minimizing environmental impact through the development and adoption of cleaner technologies, waste reduction measures, and recycling initiatives. The overarching goal is to limit adverse effects on ecosystems and prevent long-term environmental degradation.

In the context of climate change, environmental sustainability becomes crucial in addressing greenhouse gas emissions, promoting resilience to climate-related challenges, and adopting sustainable practices such as transitioning to renewable energy sources, implementing energy efficiency measures, and developing adaptation strategies [7].

Social sustainability emphasizes the importance of social equality, equal opportunities, and community engagement in cultivating a just and balanced society. It explores the creation of equitable communities where inclusivity, health, and well-being are prioritized, focusing on reducing inequality, enhancing access to essential services such as healthcare and education, and ensuring that all community members have opportunities for meaningful participation in decision-making processes. This concept also highlights the importance of cultural preservation and fostering social relations that contribute to a cohesive community identity. By promoting social justice and equity, social sustainability aims to build resilient communities capable of withstanding socioeconomic and environmental challenges, thereby securing a high quality of life for both present and future generations. However, the definition of social sustainability remains subject to diverse interpretations. Despite recent European policy emphasis on "sustainable communities" and social cohesion, there has been limited theoretical debate on precisely defining social sustainability and strategies for translating theory into real-world practices [8].

Economic sustainability entails a production system that meets current consumption levels without compromising future needs. The essence of "economic sustainability" lies in ensuring the enduring viability of the economic system itself. This concept was introduced by Hicks, who defined "income" as the amount one can consume during a period and still maintain the same well-being at the end of that period [9]. Traditionally, economists, assuming an unlimited supply of natural resources, emphasized the market's ability to allocate resources efficiently. They believed economic growth would provide the technological means to replenish depleted natural resources. However, there has been a growing realization that natural resources are finite, and the expanding economic system has placed a strain on the natural resource base. An economic system guided by the principles of "economic sustainability" adheres to the constraints of "environmental sustainability." It limits resource utilization to ensure the sustainability of natural capital and does not pursue "economic sustainability" at the expense of "environmental sustainability" [10].

The early literature on the pillars of sustainability reveals a divide between those viewing them as distinct perspectives and those adopting a systems approach. The treatment of interactions, trade-offs, and mutual reinforcements among these pillars lacks uniformity, often resulting in undertheorized integration and unnoticed value judgments, particularly concerning the economic pillar. This disagreement stems from varied interpretations and theoretical foundations within the discourse on sustainability, which is rooted in different schools of thought, contributing to confusion and competing conceptualizations. However, the inherent generality of this definition has sparked extensive reactions across academic disciplines, leading to efforts to narrow the conceptual scope of sustainable development [11]. Consequently, the lack of theoretical rigor in the discourse on sustainability and the three-pillar paradigm poses challenges in developing theory-based operational frameworks [12].

The advent of climate change represented the pivotal moment that reshaped the focus and discourse. The first Intergovernmental Panel on Climate Change [13] impact assessment report initiated a more extensive and profound contemplation of climate change, spanning various fields, sectors, and activities. Acknowledging and embracing the unavoidable magnitude, substantial risk exposures, and adaptation costs associated with climate change led to reevaluating the urgency and priorities within the sustainability discourse. It underscored the imperative for our efforts to be truly interdisciplinary and integrated. From its conception to the present day, sustainability has been extended and interpreted differently following the evolution of needs and the constraints imposed by global challenges. From this perspective, the introduction of the Sustainable Development Goals (SDGs) [14] in 2015 marked a transformative moment in the discourse on sustainable development. The 17 goals and associated targets provided a comprehensive framework that expanded the scope beyond the original three pillars to include further dimensions (Fig. 2.2).

In particular, in line with the focus of this book, Goal 11 explicitly emphasizes sustainable cities and communities, underscoring the imperative for urban areas to be inclusive, safe, resilient, and sustainable. Furthermore, climate change assumes a central role within the SDGs in Goal 13: Climate Action. This recognition has reshaped the discourse on sustainable development, accentuating the urgency of mitigating climate change impacts and fortifying resilience in the face of inevitable changes.

Fig. 2.2 Sustainable Development Goals. *Source* https://www.un.org/sustainabledevelopment/. The content of this publication has not been approved by the United Nations and does not reflect the views of the United Nations or its officials or Member States

However, despite providing targets and indicators, the UN SDGs lack a transparent and rigorous theoretical foundation and explicit descriptions of the underlying value judgments. With its complex historical origins, sustainability remains context-specific and ontologically open, necessitating an explicit understanding for rigorous operationalization [15].

The recent evolution of sustainability approaches and thinking has been significantly influenced by urbanization, emphasizing the pivotal role of cities in fostering resilience [16]. Global urbanization is an undeniable reality, poised to profoundly influence how most people live, with substantial implications for sustainable development. This centrality of urban areas aligns with the latest projections from the United Nations and UN-Habitat. The global population is expected to rise to approximately 8.5 billion by 2030, 9.7 billion by 2050, and 10.4 billion by 2100 [17]. Furthermore, urbanization is an irreversible trend; data indicates a sustained urbanization process over the next three decades, projecting an increase from 56 percent in 2021 to 68 percent in 2050 [18]. As highlighted by UN-Habitat, urban areas must anticipate and adapt to dynamic and unpredictable futures marked by various factors. These challenges underscore the necessity for cities to be equipped to respond effectively to various shocks while transitioning to sustainable, just, green, resilient, and healthy futures.

Additionally, to bolster the built environment and design role, the European Union has instituted the New European Bauhaus (NEB) [19]. This initiative aims to integrate inclusion, beauty, sustainability, and innovation to shape a more harmonious and ecological future. The fundamental objective of the NEB is to cultivate a novel lifestyle that seamlessly integrates sustainability, ensuring inclusivity and affordability for everyone.

In the discourse on the expression of sustainability in urban and environmental planning, the consequences of inconsiderate or inadequate choices in urban development decisions have been recognized as issues to be addressed [20].

Modernism has played a significant role in shaping urban landscapes, particularly in Europe and North America. However, it is important to recognize that modernist planning was not monolithic but rather a diverse set of ideas influenced by the era's scientific rationalism, which often embraced a mechanistic and deterministic view of the world. In its application to urban development, modernist thought viewed cities as an assembly of distinct parts, emphasizing technological solutions and engineered infrastructure to support urban functions. This approach also reflected a compartmentalized approach to knowledge and typically maintained a dualistic separation between humanity and the natural environment [21]. Not all modernist thinkers or planners held these views to the same extent, and significant variations and debates existed among them. However, the contrast between certain modernist principles and the integrated, holistic ethos of sustainability highlights the complexity within the modernist movement and suggests that a shift in the urban development paradigm is needed to align more closely with sustainable practices.

As the concentration of populations in urban settings escalates, offering both opportunities and challenges for sustainable development, planners need to be well-equipped to navigate these complexities [22]. Cities can promote growth, enabling economies of scale in infrastructure and the efficient delivery of social services such as education, healthcare, and governance. However, urbanization also introduces impacts like the urban heat island effect, increased impervious surfaces, and higher greenhouse gas emissions. Moreover, developing indicators and frameworks for sustainable urban development poses significant challenges in both academic research and practical applications, primarily due to the lack of a one-size-fits-all solution [23].

The ongoing discourse and research on sustainability reflect a growing awareness of the evolving dynamics, pushing the boundaries of traditional planning paradigms to pave the way for a more holistic and adaptive approach to urban design.

However, merely adhering to sustainability principles is insufficient to reclaim what has already been lost [24]. To effectively counteract climate change, a paradigm shift toward a regenerative approach becomes imperative [25, 26]. As Du Plessis explains [27]:

> While the politicians, planners, economists and businessmen were developing their versions of sustainability, the mavericks of 'radical ecologism' were opening another pathway based on a different worldview. This worldview represented a shift from seeing the planet as a deterministic clockwork system in which humans are separate from nature to seeing it as a fundamentally interconnected, complex, living and adaptive social– ecological system that is constantly in flux.

From this perspective, the limitations of a static sustainability model become apparent, emphasizing the need for a more dynamic and responsive approach to address the intricate challenges of contemporary urban environments.

As the sustainability paradigm encounters limitations in addressing contemporary urban challenges, the discourse shifts toward a regenerative approach, where sustainability is viewed as a threshold, and the regenerative approach represents the next step in the evolution of urban planning and design paradigms [28]. It is increasingly evident that five decades of pursuing sustainable development goals to reduce negative environmental impacts without impeding accepted technological and economic development patterns have fallen short [29]. Despite remarkable progress in developing tools and technologies to mitigate environmental impact, leading to the emergence of a flourishing "green" economy, these measures have not effectively slowed the inevitable decline of the global ecosystem, let alone averted the consequences of an environment rapidly deteriorating. There is a growing acknowledgment that, despite significant strides in technological innovation and policy and regulatory development, these efforts are insufficient and come too late. There is no substantial reduction in the negative impacts of consumption-based industrialized human development. It appears that the sustainable development project has proven ineffective [30].

Moving beyond sustainability, concepts like resilience, adaptation, and regeneration become imperative to effectively counteract climate change and positively impact urban environments (Fig. 2.3).

This shift is not merely a "rebranding operation" but an acknowledgment of the distinct planning paradigms represented by resilience and regenerative design, which

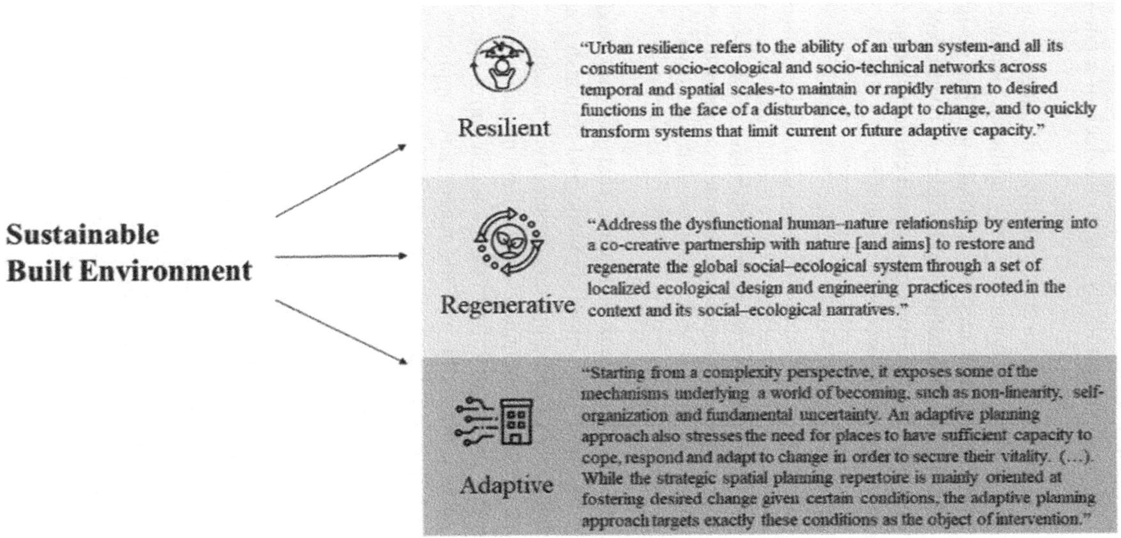

Fig. 2.3 The evolutionary planning paradigms of "sustainability." The definition of resilience is from [31]. The regenerative definition is from [25] and the adaptive definition is from [32]. Author's elaboration

share common roots with the sustainable planning paradigms but have evolved to cope with the current complex challenges [33].

2.3 The Built Environment Is a Complex System

Before proceeding with defining the planning paradigms rooted in the concept of sustainability that have evolved in the face of climate change, one must emphasize an inherent and fundamental characteristic of urban contexts: cities are complex systems. As such, they exhibit intricate interactions among their numerous components, leading to outcomes and behaviors that cannot be straightforwardly predicted from the properties of individual parts alone. Therefore, before delving further into planning paradigms that have gained relevance to cope with, mitigate, and adapt to climate change, it is necessary to take a step back to redefine the built environment and how, over the years, its conceptualization has transitioned from linear and deterministic models to recognizing it as a complex, dynamic system.

Simons defines complex systems as [34]:

> By a complex system I mean one made up of a large number of parts that interact in a nonsimple way. In such systems, the whole is more than the sum of the parts, not in an ultimate, metaphysical sense, but in the important pragmatic sense that, given the properties of the parts and the laws of their interaction, it is not a trivial matter to infer the properties of the whole.

This definition, applied to the urban context, contrasts with the theories that supported planning until the twentieth century, making a series of transformative shifts in theoretical frameworks, where the built environment is no longer viewed in isolation but as an intricate interplay of various factors, including social, economic, and ecological dynamics.

The aim of understanding the shift from linear and deterministic models to a complex model in characterizing the built environment is to better align planning strategies and interventions with contemporary urban spaces' intricate and dynamic nature. This transition signifies a departure from traditional, reductionist approaches that oversimplify the built environment, neglecting its multifaceted interactions and dependencies.

Although planners try their best to address environmental, societal, and economic issues to improve people's lives, the determinism at the base of the planning process that once a plan is drawn, it will be successful, no matter how the future unfolds, is unachievable and a heavy burden to bear. If future needs, environmental conditions, and constraints differ from those assumed during plan development, the plan is unlikely to achieve its state objectives. In the definition of something complex like the built environment, many irreducible uncertain factors (e.g., population growth, economic development, economic activities, local demands, climate change) should be considered rather than be assumed away through heroic but ultimately undefendable assumptions.

This shift is essential for several reasons. Firstly, recognizing the built environment as a complex system enables planners and designers to move beyond linear cause-and-effect relationships, embracing the non-linear, emergent, and often unpredictable behaviors inherent in urban dynamics. Secondly, it acknowledges the interconnectedness of social, economic, and ecological factors, emphasizing the need for holistic and integrated approaches to urban planning. Understanding these complex interactions is crucial for addressing contemporary challenges, such as climate change, as they are inherently linked to broader systemic issues.

Moreover, the shift toward a complex model allows for more adaptive and resilient urban interventions. Linear models often struggled to accommodate the dynamic nature of cities and were ill-equipped to respond effectively to unforeseen challenges. Embracing complexity provides a framework for anticipating and responding to urban environments' dynamic and evolving nature.

2.3.1 Shifting Perspectives: From Determinism to Embracing Complexity and Uncertainty

From the Industrial Revolution and its urban challenges, there was a growing imperative to exert control over spontaneous, disordered, and unprecedented urban configurations. Architects and urban planners sought to bring order to urban structures by employing comprehensive utopian projects characterized by rigid and dominant geometries. The concept of rational comprehensive planning, coupled with optimistic beliefs in the long-term predictability and projection of physical and social systems within cities, resulted in the dominance of deterministic models and approaches across both theoretical and practical realms of planning and design until the mid-twentieth century. Following these movements and idealized models, a revolutionary discourse in architecture and urbanism emerged: Modernism. The ideas of modern architecture and urban planning drawn up by CIAM (*Congrès Internationaux d'Architecture Moderne* 1928–1959) [35] established the core principles that shaped cities worldwide. This ideological perspective posited that urban space issues could be addressed by designing cities as machines, planning them on a city-wide scale in a top-down or centralized manner, organizing their structures hierarchically through clear geometries, and integrating the production of urban space with new industrial technologies and advancements. Urban utopias emerged as the initial deterministic models aiming to conceptualize idealized physical, social, and economic structures for urban spaces. These utopian models were precursors to deterministic and centralized urban planning and design perspectives, driven by their desire to control the entire structure through strict geometrical layouts. Even though architecture and planning concepts have radically changed their objectives, techniques, and priorities over the years, a fixed point remained "*the belief that architecture and planning are capable of transforming the world and everybody in it for the better*" [36]. Whether the vision regards a utopian city that can reform society and human behavior [37,

38], Sustainable Cities [39], or Smart City [40], there is an inherent determinism in the planning process.

Simon Richards [41] defines this standpoint as “Environmental Determinism” arguing that:

> Architect-planners are still telling people how they should be living, and this is tied to a multitude of ideas about human nature. (...) These ideas are among the most fascinating but also troubling features of the discourse of architecture and planning. They are fascinating because they sometimes involve a genuine philosophical view about the nature of humanity, about what people are and what they need. In these cases they show that architects care and think deeply about improving the conditions of life. But they can be troubling as well as they often involve passing judgment on the lives and behaviour of ordinary people, as well as a desire to reform them through the influence of the built environment.

In adherence to a technical-rational methodology, planners have sought to advance and cultivate our societal progress, constructing, and molding an envisioned physical landscape grounded in certainty, the ability to foresee our future, and, ultimately, to govern the future [42]. This outlook of an immutable world has dominated Western philosophical and scientific discourse for over 2000 years, becoming ingrained in our collective consciousness. This entrenched worldview, characterized by ideologies like modernism, functionalism, and materialism, reached its pinnacle in the twentieth century. Grounded in certainty, causality, and objectivity, this scientific paradigm undeniably yielded knowledge, understanding, and societal progress [43]. Certainty’s deep roots extended beyond the post-war period’s need for precision and control, fueled by a broader desire within social science disciplines to emulate Newtonian physics’ well-defined and straightforward nature [42]. The technical-rational approach outlined by modern architecture offered a limited angle on the multifaceted nature of cities.

Indeed, disregarding the evolutionary nature of cities over time poses a significant challenge. The limitation becomes particularly evident when examining urban challenges across extended time frames and in situations where conventional engineering practices prove inadequate in the face of uncertain, unexpected factors that can undermine the success of a plan [44, 45]. Moreover, as a complex system, the built environment responds to changes with adaptability, self-organization, resilience, and flexibility; it sets the stage for evolutionary and co-evolutionary processes [46, 46, 47]. As de Roo [42] stresses, when examining something complex like cities while excluding the temporal dimension, we observe the recurrence of traditional systems in this “fixed state,” assuming stability and balance. Essentially, the past perspectives of planners have portrayed a static reality where time played a negligible role.

In the 1990s, a significant shift occurred in the theory and practice of urban planning, challenging the prevailing belief in achieving certainty within planning processes [42]. The academic discourse about urban areas as an open complex system has challenged the technical-rational planning approach [48, 49]. The pivotal turning point came in the mid-twentieth century when critical approaches challenged the reductionist doctrines of modernism. Urban planning and design transitioned from viewing complexity as a threat to embracing it as an intrinsic feature of cities as the attempt to simplify urban areas led to sterile suburban landscapes, dysfunctional planned developments, and a failure to recognize the value of complexity in urban spaces.

Urban planners began to accept complexity as an intrinsic, natural, and inevitable feature of cities, translating this understanding into the discipline's framework. Jane Jacobs and Christopher Alexander [50, 51] played seminal roles in this shift, emphasizing that cities should be understood as problems of organized complexity rather than overly simplified entities.

This shift marked a departure from seeing complexity as a problem and conceptualizing it as an integral aspect of cities. It required planners to grapple with the notion that cities are not fixed entities but dynamic systems with self-organizing capabilities. The term "organized complexity" was suitable for characterizing the intricate relationships between various urban elements, illustrating that a controlling force does not create cities but emerges from collective interactions. In essence, the understanding of organized complexity in urban planning aligns more with the concepts of system complexity and ecological complexity rather than the complexity of objects. Cities, viewed as ecosystems, possess a dynamic, adaptable, and self-organizing nature, sharing fundamental features with ecological complexity. The term "complexity" encapsulates the higher-order phenomena stemming from the interconnected subcomponents within a system, encompassing both dynamics (processes) and structure (patterns and configurations) [52].

As described and straightforwardly summarized by Tümtürk [53], scholars have approached the conceptualization of complex systems from various perspectives. While each discipline may emphasize distinct characteristics, common features can be categorized into characteristics of system structure, system organization, and system outcomes. Specifically, (i) complex systems have a network-based interactional structure comprised of collective actions of individual components interconnected through simple interactions; (ii) Complex systems exhibit rule-based self-organization; (iii) Complex systems adapt through evolutionary processes adapting to internal and external changes; and (iv) Behaviors and outcomes of complex systems are inherently unpredictable due to the multitude of components, adaptive behaviors, and non-linearities.

Rauws [32] pointed out that cities' ability to self-adapt is a bottom-up process, where planners are not the leading entities; transformations occur, whether they are planned or not. However, that does not imply that planning has lost its function; instead, planners can become conscious guides to transition in the changing process. Nevertheless, this means embracing uncertainty. Even in seemingly straightforward issues, planning involves uncertainty. The growing internal and external complexities, coupled with the spatial and temporal dimensions, contribute to the difficulty of grasping all aspects of any project. The inability to effectively address these challenges not only jeopardizes the ongoing management of a city but also undermines its prospects for future development [54]. In addition to the intricateness inherently generated by the dynamics triggered by the many interrelated factors composing urban areas, it is necessary to account for climate change, social upheavals, and economic fluctuations, which are recognized catalysts for wicked problems [55], and deep uncertainties [56]. These uncertainties encompass unexpected natural, political, and economic events and coincidental confluences of gradual change processes leading to more significant transformations and unforeseen societal responses. This results in the pervasive reproduction of uncertainties across various domains and scales, making the future challenging, if not impossible, to predict in the long term. While planners strive to enhance the sustainability and livability of urban areas through rational interventions, the limitations of traditional planning strategies and instruments in dealing with uncertainties are evident, as they often rely on prediction, stability, and risk reduction [57].

As argued by de Roo [42], when planners encounter reality, it is unlikely that everything will be certain or uncertain; indeed, as much as cities are complex systems, there are different levels of complexity and uncertainty [58]. Planners should navigate in this nuanced area, where it is necessary to understand the problem they are dealing with to adopt the most suitable strategies.

In fact, uncertainty is a multidimensional concept Walker [59] defines as any departure from the unachievable idea of determinism. The three primary dimensions of uncertainty include location, nature, and level. All these dimensions are crucial in determining the appropriate approach to address uncertainties, particularly emphasizing the significance of the level. The levels (Fig. 2.4) are divided between determinism and total ignorance as follows [56, 59]:

- Level 1 of uncertainty. A scenario involving short-term decisions where multiple alternatives exist, each with an assigned specific probability.
- Level 2 of uncertainty. In this scenario, various alternatives are ranked by likelihood, yet no further information is available to quantify them.
- Level 3 of uncertainty, also known as deep uncertainty: In this circumstance, several alternatives can be listed, but they cannot be ranked or expressed in terms of plausibility.
- Level 4 of uncertainty. This represents the most profound stage of uncertainty, where enumerating the possibilities becomes impossible. Although assumptions can still be made, there is a risk of being wrong or surprised. Such events, beyond regular expectations, are referred to as "black swans."

Specifically, deep uncertainties are typical when dealing with a dynamic and complex system. From this classification, it is clear that planners can encounter several kinds of uncertainties, and according to the specific situation, they can choose the adequate planning strategy to cope with the uncertainties that threaten the outcome of the possible measures adopted.

In order to understand how planners have so far dealt with the different levels of complexity and, consequently, uncertainties, it is beneficial to delve into the analysis provided by de Roo, which examines the levels of complexity and describes how the planner's role is changing [42, 60] (Fig. 2.5).

Broadly speaking, the traditional system categories are snapshots of the non-linear, adaptive world that complex systems represent at the current state ($t = 0$) ranging across different levels of complexity, from two unreachable extremes, complete order

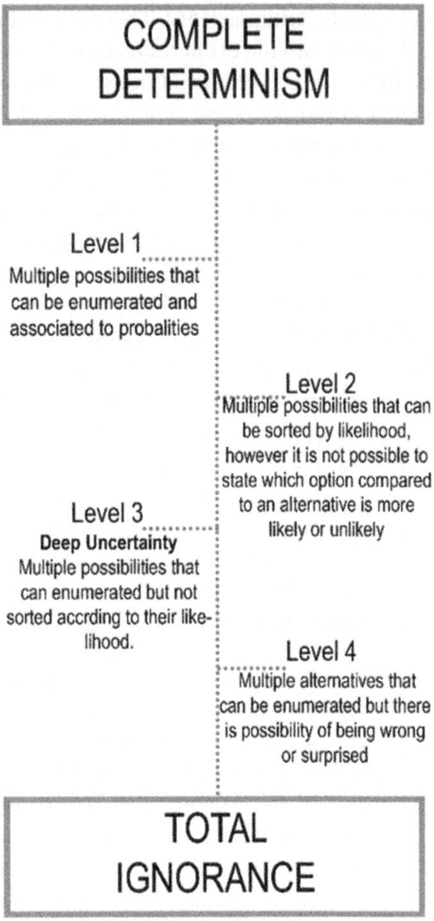

Fig. 2.4 Classification of Uncertainties' levels. Author's elaboration based on [56]

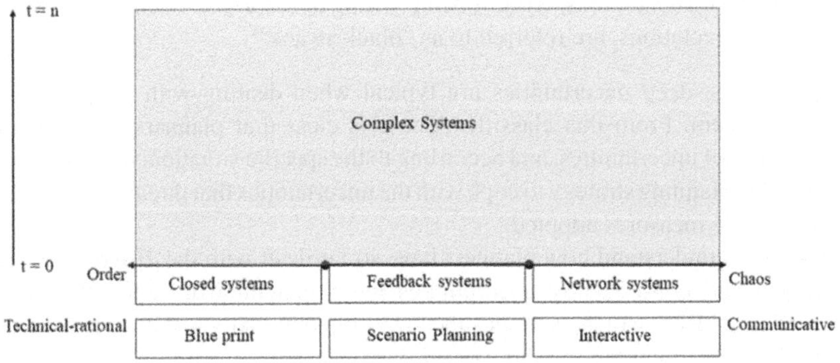

Fig. 2.5 Complex systems as described in [42]. Author's adaptation from [42]. For further details, see the text

and chaos. There are IV levels of complexity between the two extremes, relying on system theory. The levels are classified as I closed system, II feedback system, and III network system. The IV level is the complex system where we encounter the influence of time. De Roo specifies that, still, the reality is an ongoing flux that moves in time. In this frame, at $t = 0$ planners are well equipped to deal with the different a-temporal levels of complexity, with various traditional planning paradigms, from the technical-rational to the communicative [61]. Without incorporating the time factor in the analysis of complex systems, we observe these traditional systems in a "fixed state" assuming stability and balance.

Analyzing contemporary planning from this viewpoint reveals that the implicit assumption in planning activities is the "fixed state" of structure, function, fact, and value within a planning scenario.

The dynamic nature of evolving developments has been largely overlooked in the planning discourse, resulting in time being marginalized as a crucial factor. However, this is the essence of complexity thinking: acknowledging the changing processes of development and progress at various points in time.

Scenario planning is a promising aid for including time in the planning process and evaluating the effects of several levels of uncertainties, including deep uncertainties [42, 62, 63].

Dealing with deep uncertainty, particularly in the context of complex and dynamically changing urban systems, introduces the imperative of adaptability. De Roo and Rauws [60] outlined that a spatial system's adaptive capacity is intricately linked to resilience, robustness, and flexibility.

This discourse impacts and will further impact the planning procedures, the tools employed in planning, and the planner's role. In the past, planners were technicians shaping a static deterministic environment with no room for failure. Presently, planners have the challenging role of guiding the transition, where we can still rely on previous planning perspectives while exploring different tools and paradigms to encompass time, uncertainty, and complexity.

2.4 Urban Resilience, Adaptive Planning, and Urban Regeneration

In urban and architectural planning, urban resilience, adaptive planning, and urban regeneration are fundamental concepts for comprehending and tackling the challenges confronted by cities in the contemporary, rapidly changing world. These concepts gain significance as cities navigate the escalating complexity and interconnection of social, environmental, and economic issues. In the face of swift population growth, climate change, natural disasters, and socioeconomic inequalities, the imperative to prioritize sustainable urban development has reached a critical juncture.

As discussed in the previous section, the built environment is a complex system, which implies a great intricacy among the subsystems that compose the whole, generating inherent uncertainties. Moreover, recently, complexity has amplified in response to the rapidly changing climate. The challenges present a level of magnitude and intricacy unparalleled in our past experiences, necessitating an array of innovative planning and governance responses. This heightened complexity has sparked substantial discourse regarding its nature, accompanied by diverse suggestions on how to address it. While the strategies proposed vary, there is a general consensus among scholars that the unique characteristics of urban complexity, now further compounded by escalating climate uncertainty, are intricately linked to the fabric of governance and planning. Nevertheless, it becomes apparent that existing instruments of governance and planning, far from mitigating complexity, tend to exacerbate it [64].

In the quest for more effective strategies to foster a harmonious relationship between urban environments and their ecosystems, the terminologies of "sustainability," "resilience," "adaptive," and "regenerative design" emerged as evolving paradigms. However, it is crucial to recognize that these are not mere rebrandings but represent distinct design paradigms with unique goals, practices, and rules. While each paradigm has its developmental pathway, it is essential to recognize their distinctions. Sustainability seeks to maintain current development within a safe space, resilience prepares for perturbation and change, and regenerative design envisions a fundamental shift away from a dysfunctional way of living [33].

As previously addressed, an initial paradigm, sustainability, arose from recognizing the impact of human habitats on the environment and the finite nature of resources. In contrast, resilience is not a goal but a system characteristic that enables it to absorb, adapt, or transform in response to perturbations. It progresses from engineering to transformative resilience, emphasizing system adaptability and persistence. Resilience thinking acknowledges the inevitability of change and aims to manage disruptions while retaining fundamental system identity [65].

Urban adaptive design refers to a planning approach that recognizes urban development's inherent complexity and uncertainty. Originating from critiques of traditional planning methods in the 1960s and 1970s, scholars highlighted the inadequacy of rationality when dealing with what they termed as "wicked problems"—complex, fuzzy, and fluid situations. Drawing on the principles of complexity science, urban adaptive design challenges planners to embrace a situational understanding of urban development trajectories. The key idea is to guide urban transformations by creating conditions for development that strengthen a city's capacity to adapt. This contrasts with traditional planning interventions that aim to direct urban development toward a specific future configuration.

The regenerative paradigm represents a departure from the notion of achieving a net-zero impact. It calls for a fundamental shift in human roles within the global ecosystem, envisioning humans as productive members restoring and co-evolving with nature. Regenerative design represents a trajectory toward increasing wholeness and co-creative integration with nature [28].

Several definitions concerning the mentioned planning paradigms can be found in the literature, which is not a negative feature per se. However, this can generate fuzziness. Thus, the next sub-section provides an overview of "Urban Resilience,"

"Adaptive Planning," and "Urban Regeneration," analyzing where they overlap and the main discussions in the academic debate.

2.4.1 Differences and Intersections

The term "resilience" has a long and significant history, dating back to ancient times. The first scientific use of the term is credited to Francis Bacon in the 1600s [66]. Since then, resilience has been embraced in various scientific fields, including engineering, physics, ecology, management science, operations research, economics, disaster studies, urban studies, geography, sustainability science, health science, law, anthropology, psychology, and sociology [31, 67–70]. However, this interdisciplinary application of resilience has resulted in different interpretations and approaches influenced by specific norms, methodologies, and assumptions. The initial perspective focused on engineering resilience and was concerned with how quickly a system could recover its shape and size under strain. Building on this, ecological resilience emerged, focusing on the interplay between disturbances, conservation, renewal, and multiple equilibria. The concept of socio-ecological resilience then shifted the attention to the intricate relationships between humans and ecosystems. This perspective incorporated principles like self-(re)organization, adaptiveness, and learning. This evolution emphasizes the multifaceted nature of resilience, highlighting the dynamic interaction of various elements within complex systems [71].

Broadly speaking, urban resilience, as summarized by Chelleri and Baravikova [72], encompasses two primary metaphorical approaches: "*bouncing back*" and "*bouncing forward.* " These perspectives distinguish between returning to a previous stable state after a shock and embracing innovation, change, and transformation to mitigate future challenges. The shift toward the "bouncing forward" paradigm has been evident in urban resilience, integrating sustainability into disaster resilience methodologies, community perspectives, and urban planning.

Resilience thinking theories delve into the relationship between transformative and conservative perspectives, exploring incremental or disruptive innovations. In development studies, these approaches translate into "functional resilience" and "dynamic resilience," emphasizing deliberate change management and highlighting uncertainty, evolution, and non-linearity. However, implementation in cities has often followed an engineered and simplistic perspective focused on security, protection, and resistance to change, appealing to policymakers due to its alignment with existing structures.

Walker [16] discusses that resilience's "*bouncing back*" capability is often misunderstood, as its essence lies in the capacity to adapt, reorganize, and learn from disturbances without returning to the exact pre-disturbance state. Resilient systems undergo change to better handle similar challenges in the future, making them dynamic and adaptive.

Confusion arises when comparing resilience with robustness. Robustness is commonly perceived as the ability to resist disturbances without changing, akin to "engineering resilience." However, this contrasts with the dynamic nature of resilience, where adaptation is crucial. While some equate the two terms, there is a risk, especially among non-scientific policymakers, of interpreting resilience as mere robustness—resistant to change—which is understood as diminishing resilience. However, resilience and robustness should be framed in the broader context of a complex system [73].

On the one hand, resilience is a system-level concept; robustness explicitly links system dynamics to performance measures and addresses the impact of uncertainty. As a broad scientific basis, resilience encompasses self-organization, non-linearity, adaptability, and transformation. It provides a heuristic for understanding persistence and change over more extended periods, offering insights into system-level attributes. However, some scholars contend that resilience is often regarded as a buzzword, attributing its ontological ambiguity to challenges translating theoretical concepts into actionable practices [74].

On the other hand, robustness is more narrowly defined and associated with designed systems and computational methods. Robust systems maintain consistency even with incomplete or imperfect information, showing reduced sensitivity to shocks, thereby addressing the challenge of uncertainty. The distinction between robustness and resilience lies in their practical applications. Robustness tends to focus on designing fail-safe systems, ensuring consistency within a specific range, while resilience emphasizes safe-fail systems capable of learning and adapting to change. Unlike robust ones, resilient systems can evolve with change, emphasizing overlapping redundancy and the capacity to modify higher level functions. The interplay between robustness and resilience involves examining trade-offs between system stability and vulnerability. This echoes the perspective of resilience theorists, who articulate a comparable notion: transformation at one level in a system, linked to inherent fragility in a system module, is crucial for sustaining resilience at other scales within the system. Pursuing adaptability and transformation inherently carries vulnerabilities, forming an integral part of the intricate relationship between robustness and resilience.

Additionally, resilience, particularly in the context of climate change, shares common ground with adaptation, although they are distinct concepts. Despite frequent overlap in recent studies, it is essential to recognize the nuanced differences between resilience and adaptation.

The Intergovernmental Panel on Climate Change (IPCC) defines resilience as the system's capacity to anticipate, absorb, accommodate, or recover from hazardous events while preserving or improving essential structures and functions [75]. Ecosystems, evolving gradually over time, respond to climate shifts, with extreme conditions exerting substantial impacts. When an ecosystem surpasses its resilience threshold, transformation to a new state becomes probable.

The adaptive planning paradigm revolves around managing uncertainty, a fundamental aspect of the planning process. This methodology aims to confront and minimize uncertainty by continuously (re)assessing the feasibility and effectiveness of planning decisions and recognizing risks at each stage. Firstly, modeling and monitoring contribute to reducing uncertainty by enhancing scientific and professional understanding of a system. Secondly, under adaptive planning, uncertainties can guide the formulation of adaptive hypotheses, directing planning, and monitoring activities [76].

In both planning paradigms, the concept of tipping points is fundamental. Walker [16] argues that resilience involves two fundamental aspects: firstly, acquiring the ability to recognize and navigate tipping points or thresholds and, when necessary, crossing them; secondly, steering clear of undiscovered and unforeseen thresholds by gaining knowledge about the characteristics that contribute to the overall resilience of a system.

Adaptation planning explores various strategies for building resilience once tipping points in social and natural systems are met. Specifically, adaptation tipping points can be defined as critical thresholds in a system where the magnitude of change surpasses a level at which the existing strategy becomes ineffective in achieving its objectives. Beyond these tipping points, adopting an alternative adaptive strategy to address the evolving conditions and challenges becomes necessary. These points signify a crucial juncture where the current approach is no longer sufficient, prompting the need for a shift in strategies to ensure continued effectiveness in managing the system [77]. Therefore, pursuing resilience seems manageable only when dealing with a certain degree of certainty. Change is accepted, but the plan is to prepare for what we know will change. Instead, adaptive planning aims to embrace uncertainty and monitor the systems to assess when adaptation is required to steer clear of failure or seize an opportunity. Tipping points are not necessarily avoided, as alternatives are already in place to keep the system on track. In this perspective, instead of focusing on semantics, it is necessary to integrate resilience and adaptation to cope with changes we know will happen and unforeseen events.

Furthermore, resilience acknowledges the inevitability of large-scale system changes and involves developing an adaptive capacity to navigate disturbances, aiming to maintain the functional integrity of the global social-ecological system [74, 78]. It is not a static goal but a system characteristic that determines its ability to absorb and transform in response to external perturbations or internal pressures.

While resilience shares common roots with sustainability—which is about maintaining current development pathways within a safe and just operating space—resilience focuses on preparing for perturbation and change. Designing for resilience involves restructuring systems and incorporating distributed networks, diversity, and redundancy. The regenerative paradigm represents the next step. It goes beyond merely doing less harm, recognizing that real change necessitates moving away from sustaining a dysfunctional way of living and being [33].

The regenerative paradigm emerged from the realization that achieving a net-zero impact is insufficient. This paradigm shift entails reimagining the role of humans in the global ecosystem, transforming humans into productive members of the community of life. This involves restoring ecological balance, reconciling human and natural values, and creating conditions for new life through ongoing co-evolution and mutually beneficial integration of human and natural systems. The focus is on contributing to a positive impact, fostering the flourishing of life, and healing fractured systems.

This regenerative approach aligns with whole systems thinking rooted in the ecological worldview, recognizing wholeness, emphasizing the importance of relationships as a creative principle, and accepting constant change [79].

Regenerative urban design transforms the understanding of sustainability in three key ways. Firstly, it advocates for a developmental model aligned with nature's creative processes, emphasizing harmony with natural principles. Secondly, it recognizes the world as inherently unpredictable and ever-changing, reshaping the definition of sustainability to adapt to dynamic processes. Thirdly, it redefines the relationship between humans and nature, emphasizing their integration within one autopoietic system. This paradigm asserts that humans bear responsibility for the consequences of their actions and the overall health and well-being of the entire system [27].

Regenerative urban design, resilience, and adaptive urban design have emerged as promising frameworks within urban planning design and development. These paradigms depart from traditional models, emphasizing a holistic understanding of urban ecosystems, responsiveness to disturbances, and a proactive approach to sustainable development. However, despite their potential, these concepts encounter limitations in their practical applicability within the complex and dynamic context of the built environment.

Regenerative urban design, focusing on creating cities that function as ecosystems, utilizing nature's designs and processes, offers a vision of sustainable development aligned with the inherent complexities of natural systems. Similarly, resilience, acknowledging the inevitability of disruptions, emphasizes the adaptive capacity of urban systems to maintain functionality in the face of challenges. Adaptive urban design further complements these approaches by promoting flexibility and responsiveness to changing conditions.

Despite their theoretical appeal, these paradigms confront challenges when translated into real-world urban contexts.

The practical limitations arise from the complexities inherent in urban systems, where the interplay of social, economic, and environmental factors creates a dynamic and unpredictable environment. Additionally, the need for significant shifts in established urban planning practices and policies poses a considerable barrier to the widespread adoption of these progressive frameworks.

In conclusion, while regenerative urban design, resilience, and adaptive urban design offer innovative pathways for the future, their full realization faces challenges in real-world applications. Overcoming these limitations requires a nuanced understanding of the intricate urban landscape, collaborative efforts among stakeholders, and a willingness to reevaluate conventional approaches.

References

1. United Nations, United Nations conference on human environment. J. Community Dev. Soc. **4**(1), 22–28 (1973)

2. U. Wced, *Our Common Future* (Oxford University Press, Oxford, 1987)
3. D. Mebratu, Sustainability and sustainable development: historical and conceptual review. Environ. Impact Assess. Rev. **18**(6), 493–520 (1998)
4. G.A. Tanguay, J. Rajaonson, J.F. Lefebvre, P. Lanoie, Measuring the sustainability of cities: an analysis of the use of local indicators. Ecol. Indic. **10**(2), 407–418 (2010)
5. R. Goodland, The concept of environmental sustainability. Annu. Rev. Ecol. Syst. **26**(1), 251–274 (1995)
6. K. Urbaniec, H. Mikulčić, M.A. Rosen, N. Duić, A holistic approach to sustainable development of energy, water and environment systems. J. Clean. Prod. **155**, 1–11 (2017)
7. S. Riffat, R. Powell, D. Aydin, Future cities and environmental sustainability. Futur. Cities Environ. **2**, 1 (2016)
8. N. Dempsey, G. Bramley, S. Power, C. Brown, The social dimension of sustainable development: defining urban social sustainability. Sustain. Dev. **19**(5), 289–300 (2011)
9. J.R. Hicks, The foundations of welfare economics. Econ. J. **49**(196), 696–712 (1939)
10. A.D. Basiago, Economic, social, and environmental sustainability in development theory and urban planning practice. Environmentalist **19**, 145–161 (1999)
11. L.A.R. Osorio, M.O. Lobato, X.Á. Del Castillo, Debates on sustainable development: towards a holistic view of reality. Environ. Dev. Sustain. **7**(4), 501–518 (2005)
12. W.H. Clune, A.J.B. Zehnder, The evolution of sustainability models, from descriptive, to strategic, to the three pillars framework for applied solutions. Sustain. Sci. **15**(3), 1001–1006 (2020)
13. W. McG Tegart, G. Sheldon, D. Griffiths, The IPCC impact assessment: climate change (Canberra, 1990)
14. United Nations (UN), Transforming our world: the 2030 Agenda for Sustainable Development (2015)
15. B. Purvis, Y. Mao, D. Robinson, Three pillars of sustainability: in search of conceptual origins. Sustain. Sci. **14**(3), 681–695 (2019)
16. B.H. Walker, Resilience: what it is and is not. Ecol. Soc. **25**(2), 1–3 (2020)
17. United Nations (UN), *World Population Prospects 2022*, no. 9. (2022)
18. ONU Habitat, *Envisaging the Future of Cities* (2022)
19. European Commission, *Level(s) and the New European Bauhaus* (Publications Office of the European Union, 2022)
20. T. Yigitcanlar, S. Teriman, Rethinking sustainable urban development: towards an integrated planning and development process. Int. J. Environ. Sci. Technol. **12**(1), 341–352 (2015)
21. A. Heymans, J. Breadsell, G.M. Morrison, J.J. Byrne, C. Eon, Ecological urban planning and design: a systematic literature review. Sustainability **11**(13), 3723 (2019)
22. L.M.A. Bettencourt, J. Lobo, D. Helbing, C. Kühnert, G.B. West, Growth, innovation, scaling, and the pace of life in cities. Proc. Natl. Acad. Sci. U.S.A. **104**(17), 7301–7306 (2007)
23. D. Michalina, P. Mederly, H. Diefenbacher, B. Held, Sustainable urban development: a review of urban sustainability indicator frameworks. Sustainability **13**(16), 1–20 (2021)
24. T. Kadar, M. Kadar, Sustainability is not enough: towards AI supported regenerative design, in 2020 IEEE International Conference on Engineering, Technology and Innovation (ICE/ITMC), 2020, pp. 1–6
25. C. Du Plessis, P. Brandon, An ecological worldview as basis for a regenerative sustainability paradigm for the built environment. J. Clean. Prod. **109**, 53–61 (2015)
26. L.V. Gibbons, Regenerative—the new sustainable? Sustainability **1**, 1–19 (2020)
27. C. Du Plessis, Towards a regenerative paradigm for the built environment. Build. Res. Inf. **40**(1), 7–22 (2012)
28. B. Reed, Forum: shifting from 'sustainability' to regeneration. Build. Res. Inf. **35**(6), 674–680 (2007)
29. W. Steffen et al., Planetary boundaries: guiding human development on a changing planet. Science **347**(6223) (2015)
30. C. Du Plessis, Going beyond sustainability: changing views, changing ways. in *Future Challenges in Evaluating and Managing Sustainable Development in the Built Environment* (2017), pp. 40–54

31. S. Meerow, J.P. Newell, M. Stults, Defining urban resilience: a review. Landsc. Urban Plan. **147**, 38–49 (2016)
32. W. Rauws, Embracing uncertainty without abandoning planning: exploring an adaptive planning approach for guiding urban transformations. disP Plan. Rev. **53**(1), 32–45 (2017)
33. C. du Plessis, The city sustainable, resilient, regenerative—a rose by any other name? Contemp. Urban Des. Think. **Part F7**(May), 23–48 (2022)
34. H. Simon, The architecture of complexity. Proc. Am. Philos. Soc. **106**(6), 467–482 (1962)
35. E.P. Mumford, *The CIAM Discourse on Urbanism, 1928–1960* (MIT Press, 2002)
36. S. Richards, *Architect Knows Best: Environmental Determinism in Architecture Culture from 1956 to the Present* (2012), pp. 1–178
37. N.T. Dostoglu, Utopia and determinism: Architectural deterministic thinking. Metu Jfa **8**(2), 143–152 (1988)
38. M. Montavon, K. Steemers, V. Cheng, and R. Compagnon, 'La Ville Radieuse' by Le corbusier once again a case study. *PLEA 2006–23rd International Conference on Passive and Low Energy Architecture, Conference Proceedings*, September, 6–8 (2006)
39. G. Haughton, C. Hunter, *Sustainable Cities*. (Routledge, 2004)
40. A. Caragliu, C. del Bo, P. Nijkamp, Smart cities in Europe. J. Urban Technol. **18**(2), 65–82 (2011)
41. S. Richards, *Architect Knows Best: Environmental Determinism in Architecture Culture from 1956 to the Present*, 1st ed. (London, 2012)
42. G. De Roo, Being or becoming? That is the question! Confronting complexity with contemporary planning theory, in *A Planner's Encounter with Complexity*, ed. by G. de Roo, E.A. Silva (Ashgate, Farnham, 2010), pp. 19–40
43. G. De Roo, C. Yamu, C. Zuidema, *Handbook on Planning and Complexity* (Edward Elgar Publishing, Cheltenham, 2020)
44. L.M.A. Bettencourt, Impact of changing technology on the evolution of complex informational networks. Proc. IEEE **102**(12), 1878–1891 (2014)
45. M. Radhakrishnan, A. Pathirana, R. M. Ashley, B. Gersonius, C. Zevenbergen, Flexible adaptation planning for water sensitive cities. Cities. **78**, 1 (2018)
46. J. Portugali, Self-Organizing cities. Futures **29**(4), 353–380 (1997)
47. I. Carter, S. Moroni, Adaptive and anti-adaptive neighbourhoods: Investigating the relationship between individual choice and systemic adaptability. Environ. Plan. B Urban Anal. City Sci. **49**(2), 722–736 (2022)
48. P.M. Allen, *Citiesand Regions as Self-Organizing Systems: Moldels of Complexity* (Gordon and Breach Science Publishers, Amsterdam, 1994)
49. A. Faludi, Critical rationalism and planning methodology. Urban Stud. **20**(3), 265–278 (1983)
50. J. Jacobs, *The Death and Life of Great American Cities* (Knopf Doubleday Publishing Group, 1961)
51. C. Alexander, A city is not a tree. Design Council (1966)
52. G. Boeing, Measuring the complexity of urban form and design. Urban Des. Int. **23**(4), 281–292 (2018)
53. O. Tümtürk, *Designing (with) Complexity: Improving Adaptive Capacity of Urban Form By Design* (Middle East Technical University, 2018).
54. L.M.A. Bettencourt, Cities as a complex system. in *Modeling Complex Systems for Public Policies*, ed. by B.A. Furtado, P.A. Sakowski, M.H. Tovolli (Brasilia: IPEA, 2015), pp. 217–236
55. H.W.J. Rittel, M.M. Webber, Dilemmas in a general theory of planning. Policy Sci. 155–169 (1975)
56. J.H. Kwakkel, W.E. Walker, V.A.W.J. Marchau, Classifying and communicating uncertainties in model-based policy analysis. Int. J. Technol. Policy Manag. **10**(4), 299–315 (2010)
57. A.P.C.D. Milly et al., Stationarity is dead: stationarity whither water management? Science **319**(5863), 573–574 (2008)
58. G. De Roo, Knowing in uncertainty: on epistemic conditions differentiated for situations in varying degrees of uncertainty, the distinction between hierarchical and flat ontology, and the necessary merger with the axiological domain of values. disP Plan. Rev. **57**(2), 90–111 (2021)

59. W.E. Walker et al., Defining uncertainty: a conceptual basis for uncertainty management in model-based decision support. Integr. Assess. **4**(1), 5–17 (2003)
60. G. De Roo, W.S. Rauws, Positioning planning in the World of order, chaos and complexity: on perspectives, behaviour and interventions in a non-linear environment, in *Complexity Theories of Cities Have Come of Age: An Overview with Implications to Urban Planning and Design*, ed. by H. Meyer, J. Portugali, E. Stolk, E. Tan, (Springer, 2012), pp. 207–220
61. P. Healey, Planning through debate: the communicative turn in planning theory. Town Plan. Rev. **63**(2), 143–162 (1992)
62. P. Carvalho, Scenarios: learning and acting from the future. Insight Foresight (2021)
63. S. Mannucci, J. H. Kwakkel, M. Morganti, M. Ferrero, Exploring potential futures: evaluating the influence of deep uncertainties in urban planning through scenario planning—a case study in Rome, Italy. Futures **154**, September, 103265 (2023)
64. P. Allan, R. Plant, Hacking: field notes for adaptive urban planning in uncertain times. Plan. Pract. Res. **37**(6), 721–738 (2022)
65. B. Walker, D. Salt, *Resilience Thinking. Sustaining Ecosystems and People in a Changing World* (vol. 2, Island Press, Washington, D.C, 2006), p. 15
66. D.E. Alexander, Resilience and disaster risk reduction: an etymological journey. Nat. Hazards Earth Syst. Sci. **13**(11), 2707–2716 (2013)
67. D. M. Nunes, M. D. Pinheiro, A. Tomé, Does a review of urban resilience allow for the support of an evolutionary concept? J. Environ. Manage. **244**, 422–430 (2019)
68. R. Leichenko, Climate change and urban resilience. Curr. Opin. Environ. Sustain. **3**(3), 164–168 (2011)
69. A. Pathirana, M. Radhakrishnan, R. Ashley, N. H. Quan, C. Zevenbergen, Managing urban water systems with significant adaptation deficits—unified framework for secondary cities: part II—the practice. Clim. Change, 1–18, (2017)
70. R. Castaño-Rosa et al., Resilience in the built environment: key characteristics for solutions to multiple crises. Sustain. Cities Soc. **87** (2022)
71. A. Wardekker, Contrasting the framing of urban climate resilience, Sustain. Cities Soc. **75** 103258 (2021)
72. L. Chelleri, A. Baravikova, Understandings of urban resilience meanings and principles across Europe. Cities. **108** 102985 (2021)
73. J.M. Anderies, C. Folke, B. Walker, E. Ostrom, Aligning key concepts for global change policy: robustness, resilience, and sustainability. Ecol. Soc. **18**(2) (2013)
74. U. Hassler, N. Kohler, Resilience in the built environment. Build. Res. Inf. **42**(2), 119–129 (2014)
75. IPCC, *IPCC Special Report on Managing the Risks of Extreme Events and Disasters to Advance Climate Change Adaptation A Special Report of Working Groups I and II of the Intergovernmental Panel on Climate Change* (Cambridge, UK, and New York, NY, 2012)
76. S. Kato, J. Ahern, 'Learning by doing': adaptive planning as a strategy to address uncertainty in planning. J. Environ. Plan. Manag. **51**(4), 543–559 (2008)
77. J.C.J. Kwadijk et al., Using adaptation tipping points to prepare for climate change and sea level rise: a case study in the Netherlands. Wiley Interdiscip. Rev. Clim. Chang. **1**(5), 729–740 (2010)
78. J. Fehrenbacher, Resilient design: is resilience the new sustainability? inhabitat—sustainable design innovation, eco architecture, green building. Inhabitat (2012)
79. P. Mang, B. Reed, The nature of positive. Build. Res. Inf. **43**(1), 7–10 (2015)

Chapter 3
Climate Change Adaptation for the Built Environment: Addressing Urban Flood

Abstract Addressing urban flooding in the context of climate change requires a multi-scale, integrated approach that translates conceptual frameworks of sustainability, resilience, and regeneration into actionable planning strategies. This chapter delves into specific measures applied at various scales—from national policies to building-level innovations—to mitigate flooding risks and adapt urban environments to climate extremes. A critical analysis is provided on the implementation across different scales: national/regional (macroscale), city (macroscale), neighborhood (mesoscale), and building (microscale). Furthermore are discussed the challenges persisting in achieving effective communication and integration across these scales, highlighting the need for interdisciplinary collaboration and context-specific planning. By strategically incorporating adaptive measures at multiple levels, urban planners can enhance the resilience of communities, ensuring safer and more sustainable environments.

Keywords Climate change adaptation · Multi-scale approach · Nature-based solutions · Sustainable urban drainage systems · Green infrastructure · Flood mitigation · Adaptive planning · Urban design · Climate-responsive planning

3.1 Multi-scale Integrated Approach

Following the exploration of overarching planning paradigms to address climate change in Chap. 2, this chapter narrows the focus to urgent, actionable strategies within architectural and urban planning. The delineated conceptual frameworks that guide the approaches to sustainability, resilience, and regeneration are translated into planning strategies to cope with the challenges that cities face today. Among these, urban flooding emerges as a critical concern, exacerbated by the effects of intensified climate extremes.

This chapter delves into specific planning strategies applied in urban contexts to mitigate the risks associated with flooding, adapting our cities to the new reality of climate extremes. It aims to provide comprehensive insights into the planning

S. Mannucci, *Climate Adaptation in Urban Planning*,
SpringerBriefs in Architectural Design and Technology,
https://doi.org/10.1007/978-981-97-4106-9_3

measures essential for sculpting resilient and adaptive urban environments and promoting a reconnection with the natural environment.

The aim is to provide insights into the diverse planning measures essential for shaping resilient architectural and urban contexts. Through a comprehensive exploration of different scales, from national policies to building-level innovations, striving to contribute to the evolving discourse on climate-responsive planning.

Urban areas are confronted with a pressing need to reevaluate and enhance their resilience against environmental challenges, with flooding emerging as a critical concern. As climate patterns become increasingly erratic, the frequency and severity of floods are on the rise, demanding a comprehensive and adaptable approach to urban development.

In the built environment, flood protection and adaptation can be pursued at various scales with various solutions. The intricate task of tackling flood-related issues in the built environment entails the implementation of comprehensive, multi-scale strategies and interventions.

By strategically incorporating adaptation measures across different scales, communities can significantly enhance their resilience, creating a safer and secure environment [1–3]. The recognized scale of intervention is (Fig. 3.1):

- National and regional scale (macroscale)
- City scale (macroscale)
- Neighborhood scale (mesoscale)
- Building scale (microscale)

In urban environments, the mitigation of floods involves employing a combination of structural and non-structural measures. Traditional structural approaches, or "grey" solutions, involve implementing hard-engineering structures within urban stormwater drainage systems. These conventional methods, designed for quick conveyance and drainage of stormwater runoff, have proven to be both costly and inadequate in addressing the challenges posed by more frequent and intense precipitation extremes, leading to increased flooding in urban areas, a trend attributed to climate change.

Urban drainage systems are conventionally designed to accommodate runoff based on historical rain events, considering a specific duration and recurrence period (usually 1–5 years). However, the design rain does not account for changes induced by climate change observed post-system construction. Consequently, in the face of higher frequency and intensity of rain events, the drainage system, although designed according to professional standards, may struggle to effectively manage excess water resulting from more frequent flooding. The process of urbanization, marked by changes in land use, manifests as an increased proportion of impervious surfaces, leading to diminished infiltration and, consequently, a surge in the volume and speed of surface runoff. Studies indicate that a 30% increase in impermeable surfaces compared to pre-urbanization conditions results in a twofold rise in flooding over a 100-year return period.

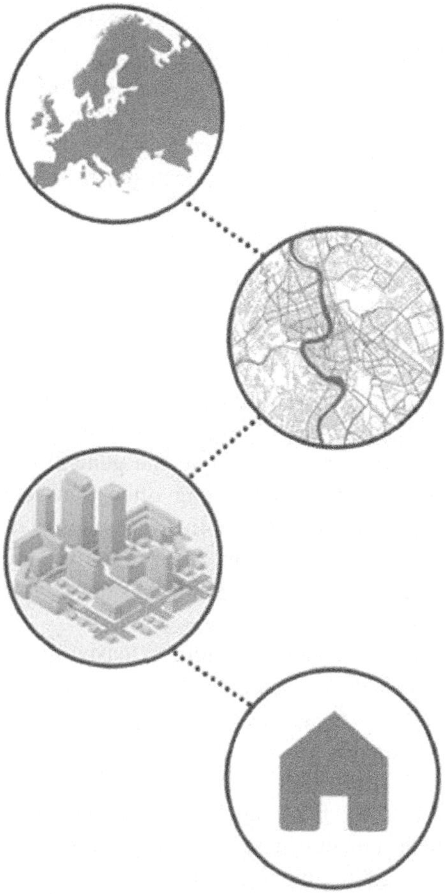

Fig. 3.1 Multi-scale approach, the different levels of intervention: National/regional scale, City scale, Neighborhood scale, Building scale. Author's elaboration

The paradigm shift from traditional urban water management to nature-based urban flood management seeks to restore pre-urbanization hydrological conditions. This involves reducing and delaying runoff by incorporating green elements that enhance infiltration, evaporation, and water retention. Cities worldwide are increasingly adopting green solutions, enacting regulations, initiating programs, and providing incentives to bolster flood protection through Nature-Based Solutions.

In this chapter, the focus shifts to the implementation of mitigative and adaptive measures for the built environment as short-term robust solutions or long-term adaptive options. Where possible, the limits of application are expressed, while other cases require context-specific assessment.

The present discussion does not delve extensively into the distinctions between structural and non-structural measures. Instead, it concentrates on the current strategies adopted mainly by planners to foster climate adaptation, leveraging regenerative, nature-based strategies and adaptive measures. The objective is to mitigate the risks and consequences associated with flooding, fostering adaptive cities in response to the escalating impacts of climate extremes and working toward a regenerative, sustainable future.

In advancing through this analysis, a comprehensive overview across all scales is provided, with a pronounced focus on the meso and micro levels. This emphasis reflects these scales' pivotal role in urban and architectural design, necessitating a deeper dive into the strategies critical for cultivating resilient and adaptive urban contexts.

3.2 Macroscale

At the national/regional scale (macroscale), collective efforts and collaboration between various cities and communities come into play. It is the scale where policies are implemented and the general management strategies are outlined [4–6]. Recognizing that flooding often transcends boundaries, these regions engage in cooperative efforts, pooling resources to develop comprehensive plans and policies for flood management. Such initiatives include conducting flood risk assessments, implementing early warning systems, and creating coordinated emergency response procedures.

At the city level, more extensive flood adaptation measures become a necessity. Such strategies involve implementing flood adaptation measures specifically designed to address the unique challenges urban environments pose. Examples include constructing robust flood barriers, using flood-proofing techniques to protect vital infrastructure, and developing elevated structures to minimize the damage caused by flooding. Localized flood maps are created, a broad framework for management is described, and a general viewpoint and methods are provided [7].

In addressing urban environmental challenges, the conventional reliance on technical and engineering approaches has been a long-standing practice [8]. However, in recent years, the growing importance of ecosystem-based approaches in urban environmental management has been recognized [9–11]. In certain instances, ecosystem-based approaches have proven to be a more effective and cost-efficient alternative than traditional engineering or technology-centric methods [12].

Among ecosystem-based elements, green infrastructure represents a pivotal concept in this evolving landscape, encompassing an interconnected network of greenspaces [13, 14]. It emphasizes the integral role of urban green space networks, positioning them not as optional but as fundamental and indispensable elements crucial for the high-quality development of urban communities [15].

Nature-based solutions represent a problem-driven and solution-oriented approach that leverages urban nature to address societal challenges and emphasizes the application of ecological wisdom to solve sustainability challenges, underlining a holistic and environmentally conscious perspective [8].

While green infrastructure emphasizes the interconnectedness of green spaces in urban planning, nature-based solutions take a broader approach by leveraging urban nature as a versatile tool to tackle specific challenges. The distinction lies in their scope, with green infrastructures emphasizing the network of greenspaces and nature-based solutions highlighting the use of urban nature as a problem-solving mechanism in a broader sustainability context.

A key concept that underlies conscious planning toward water management that relies on nature-based solutions and green infrastructures applied at the city scale is water sensitive or "Sponge City" [16, 17]. A Sponge City refers to an innovative and resilient urban design approach aimed at mitigating the adverse effects of urbanization on water resources. The concept draws inspiration from the natural functionality of sponges, emphasizing the capacity to absorb, store, and gradually release water. Sponge Cities integrate a comprehensive set of green and blue infrastructure elements to enhance water absorption, reduce surface runoff, and improve overall water resilience within urban areas. These strategies include incorporating permeable surfaces, green roofs, constructed wetlands, rain gardens, and strategically designed water retention basins. By fostering a harmonious interaction between the built environment and natural water cycles, Sponge Cities contribute to sustainable urban development, climate resilience, and the preservation of water resources. However, the implementation of such cities necessitates interdisciplinary collaboration, informed decision-making, and adaptive strategies to suit diverse geographic and climatic conditions [18–22].

In this perspective of ecological adaptive planning, flexibility is a recognized quality that improves a system's capacity to handle uncertainty [23]. For instance, the tactics used in the method outlined by Radhakrishnan et al. [24] are categorized into six groups: "retain, resist, relieve, retreat, accommodate, and prepare (4RAP)." An approach for managing flood risks holds that incorporating water into city planning is more important than defense systems that can give residents a false sense of security.

Under the umbrella of nature-based solutions, sustainable urban drainage systems (SuDs) have been demonstrated to mitigate the consequences of floods and heavy rain [25–27]. In addition, these solutions offer ancillary advantages like enhanced biodiversity, better water quality, favorable impacts on the urban microclimate, and, if properly planned, improved urban environment aesthetics, including water features that can enhance residents' overall quality of life [28, 29].

Four distinct techniques are employed to mitigate flood risk and control runoff through sustainable drainage systems [28]:

Infiltration: Excess water from precipitation or other sources seeps into the soil, where it either slowly but naturally replenishes aquifers or is used by plants to grow and then released into the atmosphere through evapotranspiration. It is imperative to take into account the properties of the soil.

Detention/Retention: Excess water is temporarily retained in areas such as ponds and basins, where it may eventually seep into the ground when adequate conditions for soil and aquifer exist, or it may be released back into the drainage system after the peak is reached.

Conveyance: Excess water is directed through systems such as swales and channels to a designated outlet or body of water.

Water Harvesting: Collected rainwater is redirected into storage solutions like reservoirs, tanks, pits, or cisterns, where it is often treated for reuse in household applications or for watering gardens.

SuDs fall under various categories which can be applied at the different scales here analyzed, with the following being the most prevalent in the built environment:

Rainfall harvesting: This technique gathers rainfall, mostly for home use, that would otherwise increase runoff.
Green roof/wall: Vegetable coverings for roofs or walls that have several advantages, including decreased energy use within the building, a decrease in the Urban Heat Island, and a longer lifespan that acts as a barrier against temperature fluctuations and ultraviolet light.
Floodable areas: Urban areas that are susceptible to flooding. These could be squares or playgrounds where water is momentarily held to relieve pressure on the drainage system after a big event.
Permeable paving: This type of paving lets water seep into the ground.
Flood park: Parks utilized as storage areas and frequently paired with other SuDs to encourage water infiltration or transportation toward a receiving feature.
Rain garden: A small soil depression that acts as an infiltration site. The garden's plants should be able to tolerate higher water levels periodically.
Basin: They are separated into basins for infiltration and detention. The first is often dry and holds onto extra water for a short while before releasing it gradually. The latter soaks up extra stormwater and lets it seep into the ground gradually.
Swales: Tiny openings that allow runoff water to enter, be stored, or be transported.
Wetlands: Ponds that are shallow and covered in water plants. They offer water purification as well as flood assistance.
Buffer zones: designated areas adjacent to water bodies or impermeable surfaces. They mitigate pollution, control stormwater runoff, and enhance water quality. These spaces, often vegetated, minimize erosion, facilitate infiltration, and reduce runoff volumes, contributing to hydrological resilience.

In the context of stormwater management at the city scale, prominent nature-based solutions include bioretention systems, buffer zones, swales, and wetlands.

Bioretention systems, or rain gardens, feature compact excavated areas filled with specialized substrates. The primary goal in designing bioretention systems is to optimize infiltration and facilitate vegetation growth. Consequently, resilient native terrestrial plants are often chosen, and their size and species are adapted to the dimensions of the bioretention system and prevailing climatic conditions. These systems are engineered to accommodate approximately 15–30 cm of runoff pooling/ponding above the substrate. Inlet structures tailored to local climatic conditions regulate runoff entry into the system, while standard features encompass underdrain and overflow pipelines interconnected with storm sewers. Overflow mechanisms are designed to redirect flows surpassing the ponding capacity. In cases where soil infiltration is undesired, impermeable bottom stormwater infiltration systems can be incorporated into the construction of bioretention cells. The capacity to diminish peak flow volumes through infiltration and evapotranspiration hinges on the soil's infiltration capacity and the system's capture volume. Consequently, bioretention systems prove efficacious in diverse settings, including urban areas, alleviating the surge in stormwater volume generated by impermeable surfaces that might otherwise overwhelm conventional drainpipes and stormwater control systems. Bioretention systems are engineered to drain within a few hours, necessitating the selection of vegetation capable of withstanding variable hydrologic conditions. Notably, these systems exhibit efficacy in peak discharge control, with various studies reporting an average reduction in peak discharge ranging from 40 to 99%. Beyond peak discharge control, bioretention is an effective strategy to curtail stormwater volumes and pollutants through natural treatment processes. However, there remains a knowledge gap concerning these systems' real-world runoff retention capacity, particularly in cold climates, given the technical intricacies associated with flow measurements in most stormwater infiltration systems [30–32].

Buffer zones can be strategically positioned on this scale between impervious surfaces like parking lots and roads and receiving watercourses like streams or ditches [33, 34]. These vegetated strips are designed to receive stormwater runoff. Their primary objective is to buffer stormwater, reducing its velocity and enhancing water quality before it enters a conveyance system. The vegetation, typically consisting of smaller plants and grass mixtures, facilitates increased infiltration while concurrently removing sediments and pollutants through sedimentation, filtration, and biological and chemical activities. Buffer zones are often implemented alongside other SuDs, like swales, to treat runoff from smaller drainage areas, such as roads or parking lots. Studies have demonstrated that vegetated buffer zones can significantly reduce inflow runoff volume, ranging from 20 to 85% when utilized as standalone measures. They facilitate water infiltration into permeable soils, contributing to effective stormwater management.

Swales contribute to improved water quality through mechanisms such as sedimentation, filtration by vegetation and plant material, and infiltration through soil. The selection of native plant species, including grasses and herbaceous species, with a specified maintenance height of 75–150 mm further enhances their effectiveness. Swales demonstrate optimal performance and are ideally suited for low-density housing areas with expansive roadway verges or open spaces where overland flows may occur. The efficacy of swales is contingent upon their size and the intensity of rainfall events. Swales typically generate no runoff during smaller rainfall events, while they act as conveyance systems during more intensive rain events, effectively delaying runoff peaks. Various studies have reported significant runoff volume reductions ranging from 23 to 48% [31].

Wetlands are perpetually wet zones designed to manage and treat runoff water through natural processes. Their classifications often consider hydrology and vegetation type, with categories including horizontal sub-surface flow, vertical sub-surface flow, free water surface (surface flow), or hybrid combining different types. Free water surfaces, commonly utilized in urban stormwater management, contribute to green spaces with recreational value. These areas typically feature emergent and submerged aquatic plants fostering biodiversity and enhancing treatment processes.

The impact of wetlands on flood mitigation varies in literature [35–37]. While some studies suggest wetlands reduce floods, others indicate no impact or even an increase. This diversity is due to the wide range of wetland ecosystems, each behaving uniquely. However, their influence is just one factor among many shaping flood magnitude, with rainfall patterns, drainage area, soil type, and vegetation cover often playing more significant roles. Notably, wetlands cover a small percentage of most river basins, and during major floods, their presence or absence may have minimal effect, as the primary determinant is usually the amount of rainfall [38].

3.3 Mesoscale

At the neighborhood level, the specific flood threat is associated with urban environments, necessitating tailored designs derived from broader strategies conceived at the city scale [1]. In water-sensitive initiatives, one prevailing approach often involves retrofitting centralized infrastructure to temporarily alleviate the impacts of extreme climatic events, addressing immediate concerns rather than embracing holistic measures. This approach falls short in adapting to unpredictable environmental conditions and evolving water risks [39].

Flood mitigation strategies at the neighborhood level increasingly emphasize nature-based solutions and sustainable drainage systems, harmonizing with the environment to promote ecological balance while efficiently managing floodwater. The strategic integration of nature-based solutions and green infrastructure networks offers a comprehensive approach to addressing social, economic, and environmental challenges, enhancing urban resilience against future threats.

Scientific research envisions merging resilience theories and adaptive management with design disciplines, emphasizing opportunities to redefine the role of urban design within a broader socio-ecological perspective.

Urban flood mitigation is identified as a primary opportunity to renew urban areas through nature-based solutions and resource co-management initiatives applicable to emerging realities. However, the current state of the art lacks clarity on how to implement an adaptive approach in urban design, and a considerable gap persists between theory and practice. The adaptive strategies required to manage the rainwater cycle from an urban design perspective remain insufficiently understood, with integrated assessment and monitoring experiments failing to articulate the links between urban water resilience objectives and project performance [39, 40].

Urban design assumes a crucial role in addressing climatic change and water hazards while revitalizing city areas. Recognizing the design process as a distinctive problem-solving tool, research underscores its role in maintaining a functional balance between human needs, environmental factors, and economic constraints. Achieving resilience raises the compelling question of how to incorporate scientific knowledge about socio-ecological systems into the urban design process and vice versa, fostering a "learning-by-doing" approach [39].

The neighborhood scale introduces important considerations for place-based approaches, community involvement, and responsive policy-making. Place-based actions that involve communities in planning and design enhance preparedness and community resilience. While acknowledging the need for local action and community-based adaptation, operationalizing neighborhood-scale design experiments has been challenging. Establishing a robust and resilient framework requires adopting a multi-scale approach focusing on each level's unique challenges and characteristics—neighborhood, city, and region. Constructing flood-resistant infrastructure and implementing non-structural measures at the community level cultivates an environment capable of withstanding and adapting to the challenges posed by flooding. Beyond configuring specific nature-based solutions, the consideration of community-based measures is paramount at the neighborhood scale. Community vulnerability, comprehensively defined, pertains to the susceptibility of a community or its individual components to the detrimental impacts of disasters. Existing variations in vulnerabilities play a pivotal role in determining the exposure of households, businesses, and communities to natural hazard consequences, significantly influencing their capacity and resources for response and recovery [41–44].

Floods do not affect people living in the same neighborhood area with the same magnitude. Their impacts differ across various community groups. Some groups may easily anticipate and respond to hazard threats, while others face more significant challenges or even impossibilities [45]. Consequently, post-disaster recovery often exhibits an uneven pattern, with specific community segments rebounding while others lag. This disparate recovery poses a threat to overall community vitality and resilience, raising concerns about its future trajectory. Therefore, alongside nature-based solutions, community-based measures are crucial in fostering a more equitable and resilient neighborhood in the face of potential disasters [46].

The discourse on spatial justice, as it intersects with flooding and climate-related challenges, has received considerable attention. Recent research has brought challenges in reconciling technical risk management frameworks with the broader political economy, especially at the neighborhood scale.

The exploration of vulnerability factors reveals that individuals and communities within neighborhoods exhibit diverse susceptibilities and responses to flooding threats. This emphasizes the unevenness in their ability to engage with and mitigate such hazards, showcasing the neighborhood scale as a critical context for understanding justice implications in climate-related challenges. Empirical studies examining disaster events and long-term stresses, such as climate change, pinpoint intricate variables contributing to vulnerability. Certain groups, stratified across economic, ethnic, social, and cultural attributes, are disproportionately exposed to hazards and less equipped to cope with their effects. For instance, extreme weather events disproportionately affect vulnerable populations within neighborhoods, posing unique challenges that demand justice-oriented solutions [47, 48].

Research indicates that despite advances in understanding flood risk nuances, policy and practice interventions aiming to reduce flood risk may inadvertently exacerbate social inequity, particularly within neighborhoods. Critiques of flood defense funding highlight potential "*winners and losers*"[48], while policy initiatives might not adequately address the specific requirements of the most vulnerable in neighborhood society. Without an appreciation of varied capacities to cope with flooding, such efforts risk perpetuating social inequities or serving the interests of the powerful and privileged within neighborhoods. This evolving discourse underscores the need for a nuanced understanding of justice and equity in the built environment within the context of climate-related challenges.

3.3.1 Strategies at the Neighborhood Scale

At the neighborhood scale, translating and applying strategies devised at the city level within the urban context necessitate a nuanced understanding of the local environment. Local adaptations play a pivotal role in ensuring the appropriateness and success of overarching measures [2, 49, 50].

Several examples of SuDs, such as green/blue infrastructures implemented at the mesoscale, are successfully aiding in coping with climate extremes [41, 51, 52]. One exemplary case is the Water Square project in Rotterdam [53] (Fig. 3.2). In 2011, the water storage system project at Benthemplein aimed to transform the space into a dynamic area for youth activities while maintaining an intimate, green, and pleasant atmosphere.

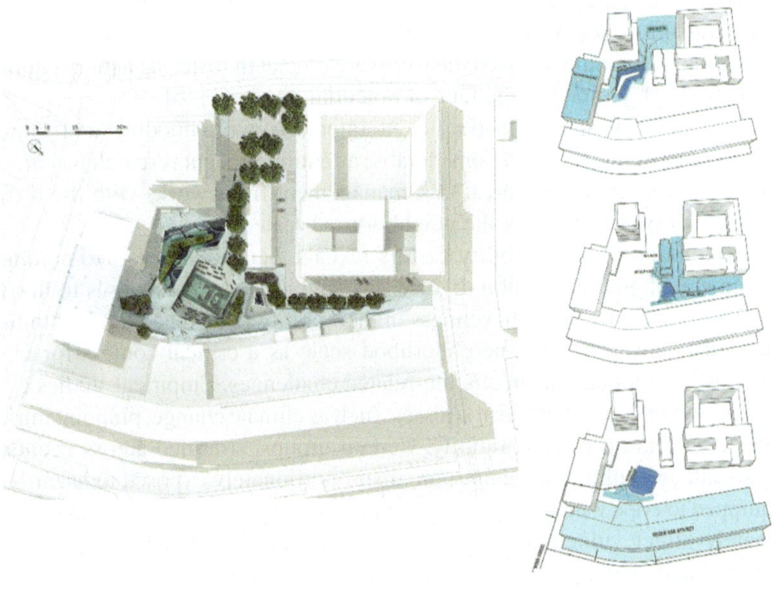

Fig. 3.2 The concept for the Benthemplein Water: the strategic filling of its basins. © The Urbanisten (www.theurbanisten.nl)

The Water Square in Rotterdam comprises three basins of varying sizes designed to collect rainwater. Two shallow basins receive rainwater whenever it rains in the vicinity. These trapezoidal basins feature low steps around the perimeter, creating a small stage in the central area.

The third basin, considerably deeper, fills with water only during consistent and intense rainfall. Once filled, water overflows from stainless steel gutters on the sides. Stylistically, areas prone to flooding are painted in blue tones, while surfaces for water flow are constructed from stainless steel.

Throughout the year, especially when not filled with rainwater, the central basin serves as an open area for sports such as basketball, football, or volleyball. Additionally, small terraced seating on the sides allows spectators to watch games. The stainless steel gutters, designed oversized for rainwater management, also serve the dual purpose of being used by skateboarders during rainy periods. This local adaptation not only addresses functional aspects like water management but also integrates recreational and aesthetic elements, showcasing the importance of tailoring strategies to the unique characteristics and needs of the neighborhood.

Another noteworthy case study for examining the intersection of climate adaptation, stormwater management, and urban regeneration is the redevelopment of Hans Tavsens Park in Copenhagen. This comprehensive project, documented through twenty-three qualitative interviews conducted between January and September 2017, showcases various stakeholders divided into two groups. The first includes individuals directly engaged in the Hans Tavsens Park project, comprising community activists, professionals, and designers. The second group involves strategic stakeholders, such as local government policy-makers and design consultants, who are participating in climate adaptation planning at the urban scale. The access to key actors involved in shaping Copenhagen as a green and resilient city enhances the robustness of the methodology.

Hans Tavsens Park's redevelopment aligns with the Copenhagen Climate Adaptation Plan, recognizing the escalating risk of stormwater flooding as a primary consequence of climate change. The Cloudburst Management Plan, initiated in response to severe flooding in 2011, diverges from traditional approaches by advocating green, surface-level water retention ponds and drainage channels. This innovative strategy aims to create "blue and green oases" and is rooted in forecasted economic benefits, including increased property values, innovation, and employment opportunities.

Hans Tavsens Park's ongoing redevelopment exemplifies the Cloudburst Plan's principles, representing a capital-intensive regeneration initiative within the framework of "Områdefornyelsen Nørrebro." This project integrates climate adaptation measures, including water retention and infiltration zones, with a redesign of the park and its surroundings by SLA, a Danish landscape architecture firm, and Ramboll. The infrastructural modifications aim to collect stormwater, channel it through conveyance channels, and transport it to a nearby lake [54].

3.4 Microscale

At the building scale, design interventions become crucial for enhancing flood resilience. This section delves into the application of design measures and technologies at the plot and building levels. From sustainable drainage systems to elevated structures, we examine how architectural innovations contribute to minimizing the impact of flooding on individual buildings and their inhabitants.

While strides are being made in flood protection and adaptation, challenges persist in fostering cohesion between different scales and disciplines. This chapter's central exploration theme is the dichotomy between specific technical solutions and a holistic, co-benefit approach. By addressing these challenges head-on, we pave the way for a more integrated, multidisciplinary, and effective approach to flood-resilient urban planning.

Technologies and design measures are implemented at the plot/building level, specifically at the building scale. The characteristics of the flood and local building codes are used to define specific actions. Integrated strategies can be created to be water- or flood-tight, and the plot can be protected with SuDs or landscaping techniques. However, if the site planning is insufficient, extra steps can be taken directly on the building to guarantee safety. Different approaches to addressing flood risk are associated with five main categories of building types (Table 3.1) [1], and they can be suitable for retrofitting or designed for new constructions.

Every approach has its own set of limitations, so it is important to think carefully about whether it makes sense to apply it for retrofitting an existing building or for new construction. One factor to take into account is the efficacy of the approach in relation to the characteristics of the flood. For pre-existing structures, the most common methods are wet- and dry-proofing.

Table 3.1 Different approaches and building types to cope with flooding and their applicability. Adapted by the author from [1]

Approach	Building type	New construction	Retrofit
Avoidance	Elevated buildings	✓	✓
Resistance	Dry-proof buildings	✓	✓
Resilient	Wet-proof buildings	✓	✓
Floating	Floating buildings	✓	×
Amphibious	Amphibious buildings	✓	×

3.4.1 Dry-Proof and Wet-Proof Buildings

In architectural design for flood mitigation, structures employing dry-proofing techniques are constructed with waterproof materials to obstruct water penetration, generally suited to regions experiencing lower levels of flooding. Conversely, buildings adhering to wet-proofing principles are designed to endure water entry in a controlled manner, mitigating structural damage and enabling a swift post-flood recovery and reoccupation.

One of the advantages of wet-proofing lies in its ability to equalize water levels inside and outside the building quickly, significantly diminishing the impact of the hydrostatic pressure. This reduction in pressure can substantially lower the risk of structural damage during flooding.

Effective wet floodproofing strategies include: (i) Facilitating the unobstructed entry and discharge of floodwaters in and out of the building. (ii) Minimizing damage to parts of the building below the flood level through exposure to floodwaters. (iii) Safeguarding utility and mechanical systems inside and outside the building.

Dry-proofing keeps water out during a flood [55]. It is usually applied to water levels less than 0.6 m. The structure should be tested and strengthened to withstand hydrostatic pressure and provide additional isolation, such as through waterproof membranes. Utilizing closed-cell insulation is also recommended. Water-resistant finishing materials like concrete, bricks, and flood-resistant steel or aluminum windows and doors are also required. The building's speedy recovery is the goal, and the technology used may differ from that required for construction modifications. Still, wet-proofing is suitable for flood levels starting at 0.6 m and can be used with dry-proofing to guarantee a rapid recovery,

3.4.2 Elevated Buildings

During a flood disaster, the most common approach to preventing damage and maintaining normal life is to employ elevated buildings. Elevated structures come in a variety of forms; they can be placed on structural walls, constructed with steel and reinforced concrete, or supported by posts built from local materials. Building design should be described considering the context's characteristics and flood hazard. Elevated buildings are not flexible, so the maximum water level should be considered for various entrances to provide safe access. This kind of construction is suitable for varying water depths.

3.4.3 Floating Buildings

Floating buildings (Figs. 3.3 and 3.4) have a light structure, usually made with timber or steel, resting on a buoyant concrete foundation.

They are anchored to two mooring posts where the building slides following the water level; they are placed on the diagonal to avoid overturning.

Archimedes' concept is the foundation for buoyancy, and the same computation that applies to boats also ensures stability. As a result, the metacenter, the center of buoyancy, and the center of gravity must all line up. Since the building's design cannot be altered after it is decided, the weight distribution is computed to ensure stability. Typically, floating buildings are constructed at a factory and then delivered

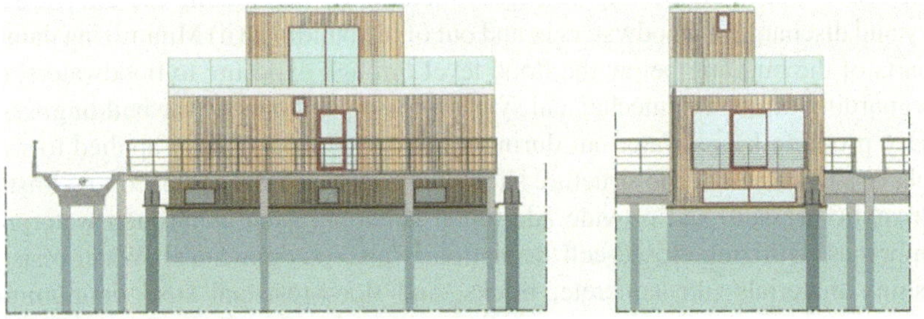

Fig. 3.3 Floating buildings elevations. Author's elaboration

Fig. 3.4 Floating building section. Author's elaboration

Fig. 3.5 Ijburg floating homes, on the left the houses built on public commission. On the right the houses built on private commission. (Photos taken by the author)

to the location. Jetties, which also connect utility lines that are too large to follow the vertical excursion, connect them to the mainland. The sleeping quarters and a utility room are below water level, and the concrete hull is intended to be a part of the living area. A utility room is required because a lifting pump needs to be placed in the lowest area of the building.

An exemplar of successful water-integrated urban planning is IJburg in Amsterdam, which boasts Europe's largest floating residential district, [56] (Fig. 3.5) It is a neighborhood, that is surrounded by water, and divided into an eastward-realized private commission area and a public area. The project includes 110 private plots, 55 public houses, and 165 floating homes.

The neighborhood, with a home density mirroring central Amsterdam's iconic districts, features floating residences accessible via public jetties, emphasizing its integration with the city rather than isolation. Homes are distinctly set apart by water, with service connections cleverly concealed within the jetty structure, accommodating the water level fluctuations. Material choices favor environmental harmony and maintenance ease, with light-colored facades and a blend of white and brown hues. Stability and functionality are core to the design, with a focus on internal balance and strategic placement of heavier fixtures to ensure comfort and practicality within these innovative living spaces.

3.4.4 Amphibious Buildings

In the event of a flood, an amphibious building (Figs. 3.6 and 3.7) employs floating technology but remains typically on the ground. It is held in place by guideposts where the structure rises and lowers in response to the water level, which must be higher than one meter to guarantee buoyancy. A wet dock made of concrete is set next to the building's concrete caisson. Several examples from worldwide [57–59] attest to the efficacy of this technology. The foundation of an amphibious building is the same as that of the preceding example.

The amphibious homes have a concrete hull with a lightweight structure built on top, and unlike a floating home, it can be built on-site. As Barker and Coutts

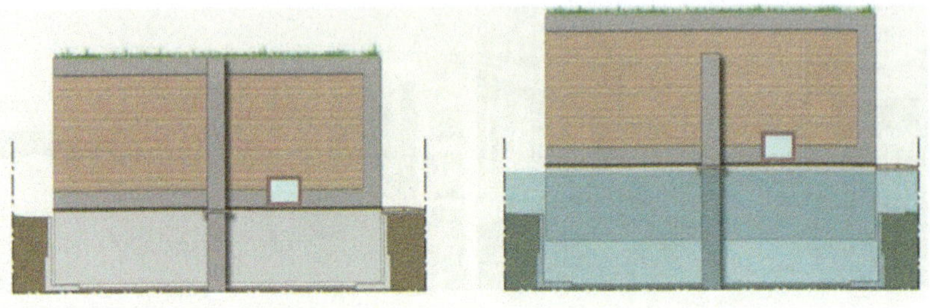

Fig. 3.6 Amphibious house elevations in the two different configurations, resting on the ground and floating. Author's elaboration

Fig. 3.7 Amphibious house section. Author's elaboration

[1] stated, the building needs maintenance because it cannot float for several years. Therefore, periodic testing is done once a year to ensure floating. The utilities must have flexible, oversized cables to handle the vertical excursion.

For this kind of test, the houses are raised to a height of 50 cm while water is pumped into the wet docks. High water variation can be handled by amphibious structures. However, they are susceptible to strong currents, just like floating buildings.

In conclusion, every situation is unique when it comes to buildings, and while some building types can be retrofitted with a combination of techniques (Fig. 3.8), some types are only intended for use in new construction due to the requirements that must be met to ensure flood safety. Table 3.2 provide a summary of general guidelines based on flood characteristics (depth and velocity).

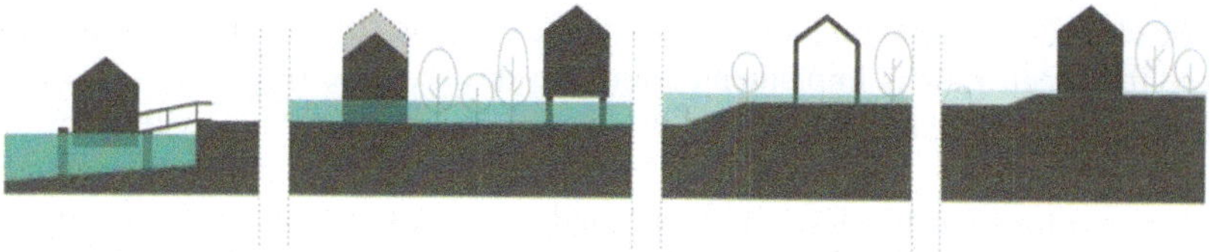

Fig. 3.8 Profile with different building types, from the left: floating, amphibious, elevated, wetproof, and dry-proof, and the associated water depths

Table 3.2 Different building types and their applicability related to flood characteristics. Adapted by the author from [1]

	Flood characteristics						
	Depth (m)				Velocity (m/s)		
Building type	<0.3	0.3–0.6	0.6–2	>2	<1	1–2	>2
Dry-proof	✓	✓	×	×	✓	×	×
Wet-proof	✓	✓	✓	×	✓	×	×
Elevated	✓	✓	✓	✓	✓	✓	✓
Floating	×	×	×	✓	✓	×	×
Amphibious	×	×	✓	✓	✓	✓	×

3.4.5 *Sustainable Drainage Systems at the Building Scale*

At the microscale, sustainable urban drainage systems (SuDs) play a pivotal role in managing stormwater runoff and enhancing environmental sustainability. Among the diverse SuDs solutions, green roofs and green walls have emerged as prominent strategies for mitigating the impact of urban development on local water systems. These innovative approaches contribute to stormwater management and offer additional benefits such as improved air quality, temperature regulation, and enhanced aesthetic value.

In this context, a closer examination of green roofs and green walls reveals their effectiveness in promoting microscale sustainability and regeneration.

Green roofs offer valuable ecosystem services to urban centers; they have been proven as an effective aid in reducing runoff, mitigating urban heat islands for urban contexts, improving thermal comfort, acoustic insulations, and air quality, and contributing to the urban landscape [47, 60–67].

As Leite and Antunes [61] describe, in contemporary construction practices, green roofs are structured with distinct layers comprising plants, substrate, filter, and drainage components. The design and selection of materials play a pivotal role in the construction of green roofs, directly influencing hydrological and thermal performance and the system's lifespan.

The choice of vegetation must align with the climatic characteristics of the region where the green roof is implemented and the desired type of growing. Additionally, incorporating features like food production, such as growing vegetables, may escalate green roof expenses due to increased maintenance requirements, including regular irrigation, fertilization, replanting, and substrate replacement. Green roofs are typically categorized based on the thickness of the growing medium allocated for each planting type. The primary classifications include extensive, semi-intensive, and intensive green roof typologies. In broad terms, semi-intensive and intensive green roofs require a substrate depth exceeding 150 mm to accommodate medium to large species, such as aromatic herbs, shrubs, and small trees. In contrast, extensive green roofs maintain a substrate depth below 150 mm, suitable for cultivating grasses, succulents, and other small plants. Extensive roofs exhibit lower hydrological performance compared to intensive roofs, resulting in the retention of a lesser amount of rainwater. Nevertheless, the advantages of extensive roofs lie in their lower construction costs [65].

While establishing threshold values for the applicability of building-related measures often involves clear parameters, it becomes apparent that such values are inherently context-specific for green roofs. The effectiveness of green roofs in managing runoff, influenced by factors like substrate composition, plant selection, and local climate conditions, highlights the need for a nuanced, site-specific approach rather than universally applicable thresholds. In various climatic zones, studies on green roofs have shown that flat green roofs retain about 94.9% of runoff, while those with a 15° slope retain around 84.9% [61]. Factors such as substrate depth, plant type, and rainfall intensity significantly affect retention.

In Kõiv-Vainik et al. [31], the authors provide an efficient overview of green roofs' effectiveness in retaining rainfall depending on the storm event's magnitude and duration, alongside other influencing factors. Green roofs can retain 88% of small storms (<25.4 mm), over 54% for medium storms (25.4–76.2 mm), and 48% for large storms (>76.2 mm). They effectively delay small, short rainfall events for 1– 4.5 h. Efficient retention occurs when the substrate layer is dry after a rainless period, but green roofs can retain moderate rainfall even on a wet substrate.

Different studies highlight the impact of substrate thickness on retention. Increasing thickness generally enhances retention, but the practicality and cost implications need consideration. Plant choice, including succulents and grass species, is crucial. Succulents tend to retain more in dry conditions, while grasses exhibit consistent retention due to higher evapotranspiration rates. Climate-specific considerations are essential. For instance, in tropical zones, succulents may lose water more rapidly, impacting retention, while in temperate zones, different species and substrates influence performance. Studies emphasize that green roofs' hydrological efficiency evolves, with factors like substrate aging influencing retention. To mitigate construction challenges, additives such as biochar and hydrogel are suggested to enhance retention without increasing substrate depth [68–72].

Transitioning from the discussion on green roofs, it is pivotal to explore another innovative green infrastructure element: green walls. The term "green wall" encompasses diverse vegetated wall surfaces, constituting a vertical bioretention system installed on facades or walls, thus overcoming spatial constraints associated with conventional setups. This conceptual alignment with traditional bioretention systems for urban runoff control. The foundational concept of green walls is rooted in the belief that nature can provide solutions to the challenges posed by urbanization. Despite extensive research on the environmental and economic benefits of green roofs, such as water conservation, stormwater management, aesthetic appeal, and habitat provision for wildlife, the potential of green walls, particularly in stormwater management, remains underexplored.

As assessed by Manso and Castro-Gomes [73], selecting appropriate vegetation for green walls is a nuanced process intricately influenced by climatic conditions, building characteristics, and the surrounding environment. Typically, Green Wall systems fall into two main categories: Green Facades and Living Walls.

Green Facades are characterized by their simplicity, with minimal technological components. In this setup, plants can either directly grow on the wall or climb an indirect element alongside it. This approach, known for its easy installation, relies on vegetation to naturally cover the building surface [74]. The selection of specific plant varieties is crucial for Green Facades, accommodating heights up to 25 m. Achieving complete wall coverage may take several years, and the foliage can be evergreen or deciduous. Depending on the supporting system, Green Facades can be further classified into Direct Green Facades or Indirect Green Facades [75].

Living Walls, on the other hand, involve techniques where vegetation grows on dedicated supporting elements, requiring regular irrigation. The vegetation uniformly covers the wall surface. Like Green Facades, living walls are categorized into Continuous Living Walls (CLW) and Modular Living Walls (MLW). CLW utilizes permeable screens, such as foam or felt layers, as growing media, primarily employing the hydroponic method [73]. Conversely, MLW employs supporting elements like trays, vessels, flexible bags, or planter tiles containing growing media and plants [73].

Green walls have demonstrated their effectiveness across various dimensions, showcasing their versatility beyond ecological advantages, becoming a valid asset to cope with multiple climate-related issues in the built environment as they hold several co-benefits.

In addition to their aesthetic contributions, vertical green systems are efficient insulators for buildings. The plant media attached to walls acts as a natural barrier, enhancing insulation and contributing to energy efficiency. This insulation property aids in temperature regulation, reducing the reliance on heating or cooling systems and promoting sustainable building practices. CLW excels in improving thermal comfort within interiors [76, 77]. Their insulating effect helps moderate indoor temperatures, creating a more comfortable and balanced atmosphere for occupants. This enhancement in thermal comfort adds to the overall well-being of individuals within the built environment [78].

Furthermore, green walls offer noteworthy acoustic performance. The plant media absorbs and dampens sound as a natural sound barrier. This acoustic insulation capability contributes to a quieter and more serene indoor environment, which is particularly beneficial in urban settings with high noise levels [79, 80]. As regards floods, they can contribute to the mitigation of runoff. Studies indicate that these systems have the potential to retain substantial amounts of rainfall, ranging from 45 to 75%. They hold promise as effective stormwater management solutions, leveraging vertical planting to capitalize on wall areas, which often surpass roof areas, especially within urban development zones. The collected stormwater can be repurposed for irrigating the green wall, enhancing on-site infiltration and evapotranspiration. However, it is noteworthy that some reports suggest that while green walls offer significant benefits, they might not surpass the performance achieved by green roofs [81–83].

The design and implementation of vertical green systems for runoff control necessitate meticulous planning, considering various factors such as structural integrity, efficient drainage systems, appropriate material selection, advanced construction techniques, and aesthetic considerations. Each of these factors plays a pivotal role in ensuring that the green wall can seamlessly accommodate vegetation, effectively mitigate water buildup, withstand challenging weather conditions, adhere to pertinent regulations, and enhance the overall ambiance of the surrounding environment. By consistently prioritizing and giving careful attention to these crucial elements, developing a highly efficient, visually captivating, and environmentally sustainable green wall becomes entirely plausible. Such a structure adeptly manages runoff and contributes significantly to the intrinsic beauty and enduring viability of the space it occupies [84].

The efficacy of both green roofs and vertical green systems as nature-based solutions at the building scale for mitigating various climate-related issues has been well-established. However, it is crucial to acknowledge the influential role of location and climatic conditions in determining their effectiveness. Given this context, employing model-based techniques to systematically analyze case-specific factors and discern the tipping points for successfully implementing these strategies is essential. A nuanced understanding can be developed through meticulous investigation using such methodologies, providing insights into the optimal conditions and thresholds that maximize the benefits of green roofs and vertical green systems in diverse environmental contexts.

3.5 Discussion and Future Outlooks

The existing body of research on climate change adaptation in the built environment has undeniably grown, emphasizing the significance of resilience and adaptive planning. However, notable limitations and gaps persist within this expansive literature, demanding critical attention.

Adopting a multi-scale strategy for flood management, especially within the framework of adapting to climate change, faces several challenges that can detract from its success. These challenges stem from the difficulty of precisely forecasting and controlling flood risks amidst climate variability in the long term; despite the recognized necessity of integrating all scales, there is a gap concerning the complexities involved in executing multi-scale applicative framework and modeling that actually bridges all the scales.

Furthermore, currently, the literature focuses on specific strategies that need to be addressed from a local perspective, evaluating the context-specific best strategy. While the planning strategies showcase a certain degree of flexibility toward adaptation, it is still unclear how and when to adapt to the built environment, especially in the case of unforeseen long-term events.

Despite the multitude of studies available, there is a tendency to concentrate on specific facets of adaptation, often neglecting the broader systemic challenges that play a pivotal role. This critical gap necessitates urgent consideration to cultivate a more comprehensive perspective that acknowledges the intricate interplay of various factors.

Furthermore, a substantial dearth of research explicitly addressing the practical implementation of long-term adaptation strategies in real-world scenarios is evident. While theoretical frameworks and conceptual models provide valuable insights, the transition from theory to practical solutions remains intricate and warrants thorough investigation. Bridging this gap is imperative to effectively combat the multifaceted impacts of climate change on the built environment.

An inclusive approach, integrating diverse disciplines like architecture, engineering, environmental science, and social sciences, is crucial for developing a nuanced understanding of long-term strategies. Each discipline contributes unique perspectives, methodologies, and expertise, collectively enhancing our comprehension of the complex challenges posed by climate change. Currently, the interaction between the experts is limited, and there are no integrated tools or methodologies to facilitate the dialogue among professionals from different fields. This interdisciplinary collaboration is pivotal in addressing the current limitations and fostering a more holistic approach to adaptation planning.

As discussed in the previous chapter, the built environment is complex; therefore, there are no easy solutions for complex issues. In the context of flood adaptation in anthropized environments, embracing a holistic multi-objective methodology is essential. Key aims, including economic feasibility, long-standing resilience, and environmental sustainability and regeneration, should be carefully weighed during the planning and implementation stages. Moreover, while we cannot stop planning for our cities in the short term, applying the most suitable strategy for the given context, especially concerning flood adaptation, planners must acknowledge the inherent uncertainties and dynamic nature of long-term changes. In the short to mid-term, it remains imperative to identify and implement the best planning strategies tailored to specific contexts. Economic feasibility, resilience, and environmental sustainability should continue to guide decision-making during these phases.

However, the planning process should not conclude with the immediate goals in mind. Instead, it should incorporate a forward-looking perspective that acknowledges the evolving nature of environmental, social, and economic conditions. Uncertainties, can significantly impact the effectiveness of chosen strategies over the long term.

Therefore, planners must integrate continuous evaluation and adaptation mechanisms into the planning framework. This involves assessing how uncertainties might affect the chosen strategy in the long term and developing alternative routes to navigate potential challenges. Flexibility and adaptability should be ingrained in the planning process, allowing for adjustments based on emerging data, technological advancements, and shifts in societal needs.

In essence, while short to mid-term plans demand a focus on the current best strategies, the planning process should not operate in isolation from the potential long-term implications. By embracing a dynamic and adaptive approach, planners can ensure the continued success and resilience of the built environment in the face of evolving challenges and uncertainties.

The next chapter will introduce promising methodologies and tools that can be integrated into architectural and urban planning to incorporate long-term perspectives.

References

1. R. Barker, R. Coutts, *Aquatecture: Buildings and Cities Designed to Live and Work with Water* (RIBA Publishing, London, 2016)
2. D. Serre, B. Barroca, M. Balsells, V. Becue, Contributing to urban resilience to floods with neighbourhood design: the case of Am Sandtorkai/Dalmannkai in Hamburg. J. Flood Risk Manag. **11**, S69–S83 (2018)
3. C. Zevenbergen, W. Veerbeek, B. Gersonius, S. Van Herk, Challenges in urban flood management: travelling across spatial and temporal scales. J. Flood Risk Manag. **1**, 81–88 (2008)
4. S. Begum, M.J. Stive, J.W. Hall, *Flood Risk Management in Europe* (2007)
5. H. De Bruijn, M. De Bruijne, E. Ten Heuvelhof, The politics of resilience in the Dutch 'Room for the River'-project. Proc. Comput. Sci. **44**(C), 659–668 (2015)
6. M. Woolgar, Making space for water. Int. Water Power Dam Constr. **60**(11), 42–43 (2008)
7. M. Balsells, B. Barroca, V. Becue, D. Serre, Making urban flood resilience more operational: current practice. Proc. Inst. Civ. Eng. Water Manag. **168**(2), 57–65 (2015)
8. J. Kim, J. Kang, Development of hazard capacity factor design model for net-zero: evaluation of the flood adaptation effects considering green—gray infrastructure interaction. Sustain. Cities Soc. **96** 104625 (2023)
9. F. Marando, E. Salvatori, A. Sebastiani, L. Fusaro, F. Manes, Regulating ecosystem services and green infrastructure: assessment of urban heat island effect mitigation in the municipality of Rome, Italy. Ecol. Modell. **392**, 92–102 (2019)
10. European Commission, *Green Infrastructure (GI)—Enhancing Europe's Natural Capital* (2013)
11. S. Narayan et al., The value of coastal wetlands for flood damage reduction in the northeastern USA. Sci. Rep. **7**(1), 1–12 (2017)
12. B.L. Keeler et al., Social-ecological and technological factors moderate the value of urban nature. Nat. Sustain. **2**(1), 29–38 (2019)
13. X. Fang, J. Li, Q. Ma, Integrating green infrastructure, ecosystem services and nature-based solutions for urban sustainability: a comprehensive literature review. Sustain. Cities Soc. **98**, 104843 (2023)
14. C. Bartesaghi Koc, P. Osmond, A. Peters, Towards a comprehensive green infrastructure typology: a systematic review of approaches, methods and typologies. Urban Ecosyst. **20**(1), 15–35 (2017)
15. S. Ronchi, A. Arcidiacono, L. Pogliani, Integrating green infrastructure into spatial planning regulations to improve the performance of urban ecosystems. Insights from an Italian case study. Sustain. Cities Soc. **53**, 101907 (2020)

16. T.T. Nguyen, H.H. Ngo, W. Guo, X.C. Wang, A new model framework for sponge city implementation: emerging challenges and future developments, J. Environ. Manage. **253**, 109689 (2020)
17. B. Gersonius, R. Ashley, C. Salinas-Rodríguez, J. Rijke, M. Radhakrishnan, C. Zevenbergen, *Flood Resilience in Water Sensitive Cities: Guidance for Enhancing Flood Resilience in the Context of an Australian Water Sensitive City* (2016), pp. 1–77
18. Y. Jiang, C. Zevenbergen, Y. Ma, Urban pluvial flooding and stormwater management: a contemporary review of China's challenges and 'sponge cities' strategy. Environ. Sci. Policy **80**(December), 132–143 (2018)
19. B.J. He, J. Zhu, D.X. Zhao, Z.H. Gou, J. Da Qi, J. Wang, Co-benefits approach: opportunities for implementing sponge city and urban heat island mitigation. Land Use Policy **86**(May), 147–157 (2019)
20. F.K.S. Chan et al., 'Sponge City' in China—a breakthrough of planning and flood risk management in the urban context. Land Use Policy **76**(March), 772–778 (2018)
21. M. Kuller, P.M. Bach, D. Ramirez-lovering, A. Deletic, Framing water sensitive urban design as part of the urban form: a critical review of tools for best planning practice. Environ. Model Softw. **96**, 265–282 (2017)
22. C.N.A. Salinas Rodriguez, R. Ashley, B. Gersonius, J. Rijke, A. Pathirana, C. Zevenbergen, Incorporation and application of resilience in the context of water-sensitive urban design: linking european and australian perspectives. Wiley Inter. Discip. Rev. Water. **1**(2), 173–186 (2014)
23. M. Radhakrishnan, A. Pathirana, R.M. Ashley, B. Gersonius, C. Zevenbergen, Flexible adaptation planning for water sensitive cities. Cities. **78**(March), 1 (2018)
24. M. Radhakrishnan et al., Context specific adaptation grammars for climate adaptation in urban areas. Environ. Model. Softw. **102**, 73–83 (2018)
25. S. Kok, A. Bisaro, M. de Bel, J. Hinkel, L.M. Bouwer, The potential of nature-based flood defences to leverage public investment in coastal adaptation: cases from the Netherlands, Indonesia and Georgia. Ecol. Econ. **179**, 106828 (2021)
26. E. Schroll, J. Lambrinos, T. Righetti, D. Sandrock, The role of vegetation in regulating stormwater runoff from green roofs in a winter rainfall climate. Ecol. Eng. **37**(4), 595–600 (2011)
27. M. Zandersen, J.S. Oddershede, A.B. Pedersen, H.Ø. Nielsen, M. Termansen, Nature based solutions for climate adaptation—paying farmers for flood control. Ecol. Econ. **179**, 106705 (2021)
28. J. Griffiths, Sustainable Urban Drainage. In *Reference Module in Earth Systems and Environmental Sciences* (2016)
29. F. Rosso, S. Mannucci, M. Morganti, S. Mariani, C. Cecere, M. Ferrero, The effect of sustainable urban drainage systems on outdoor comfort and runoff. J. Phys. Conf. Ser. **1343**(1) (2019)
30. A. Roy-Poirier, P. Champagne, Y. Filion, Review of bioretention system research and design: past, present, and future. J. Environ. Eng. **136**(9), 878–889 (2010)
31. M. Kõiv-Vainik et al., Urban stormwater retention capacity of nature-based solutions at different climatic conditions. Nature-Based Solut. **2**, 100038 (2022)
32. A.P. Davis, W.F. Hunt, R.G. Traver, M. Clar, Bioretention technology: overview of current practice and future needs. J. Environ. Eng. **135**(3), 109–117 (2009)
33. D.L. Correll, Principles of planning and establishment of buffer zones. Ecol. Eng. **24**(5), 433–439 (2005)
34. D. Mullan, K. Vandaele, J. Boardman, J. Meneely, L.H. Crossley, Modelling the effectiveness of grass buffer strips in managing muddy floods under a changing climate. Geomorphology **270**, 102–120 (2016)
35. S.D. Lloyd, T.H.F. Wong, B. Porter, The planning and construction of an urban stormwater management scheme. Water Sci. Technol. **45**(7), 1–10 (2002)
36. W.J. Mitsch, J.G. Gossilink, The value of wetlands: importance of scale and landscape setting. Ecol. Econ. **35**(1), 25–33 (2000)

37. G. Padmanabhan, M.L. Bengtson, Assessing the influence of wetlands on flooding. Wetlands Eng. River Restor. 1–12 (2001)
38. M. Acreman, J. Holden, How wetlands affect floods. Wetlands **33**(5), 773–786 (2013)
39. E. Palazzo, From water sensitive to floodable: defining adaptive urban design for water resilient cities. J. Urban Des. **24**(1), 137–157 (2019)
40. S.T.A. Pickett, M.L. Cadenasso, J.M. Grove, Resilient cities: meaning, models, and metaphor for integrating the ecological, socio-economic, and planning realms. Landsc. Urban Plan. **69**(4), 369–384 (2004)
41. A. Raffa, Design urban climate-resilience with nature-based solutions and green infrastructures. Challenges, issues, and best pratices for neighborhood-scale regeneration. Anu. d'Arquitectura i Soc. Res. J. **3** 234–266 (2023)
42. A. Bukvic, A. Borate, S. Hughes, R. Weaver, D. Imburgia, W.A. Stiles, Exploring neighborhood-level resilience to flooding: why the context and scale matter. J. Flood Risk Manag. **14**(2), 1–15 (2021)
43. M. Zellner, D. Massey, E. Minor, M. Gonzalez-Meler, Exploring the effects of green infrastructure placement on neighborhood-level flooding via spatially explicit simulations. Comput. Environ. Urban Syst. **59**, 116–128 (2016)
44. J. Kim, Exploring green infrastructure benefits at house and neighborhood scale: case study of Illinois, USA. Landsc. Ecol. Eng. **14**(1), 165–174 (2018)
45. N. Mohtat, L. Khirfan, Distributive justice and urban form adaptation to flooding risks: spatial analysis to identify Toronto's priority neighborhoods. Front. Sustain. Cities **4** (2022)
46. S. van Zandt, W.G. Peacock, D.W. Henry, H. Grover, W.E. Highfield, S.D. Brody, Mapping social vulnerability to enhance housing and neighborhood resilience. Hous. Policy Debate. **22**(1), 29–55 (2012)
47. S. Mannucci, F. Rosso, A. D'Amico, G. Bernardini, M. Morganti, Flood resilience and adaptation in the built environment: how far along are we? Sustainability **14**(4096), 1–22 (2022)
48. P. O'Hare, I. White, Beyond 'just' flood risk management: the potential for—and limits to—alleviating flood disadvantage. Reg. Environ. Chang. **18**(2), 385–396 (2018)
49. R. Ewing, S. Hamidi, Compactness versus sprawl: a review of recent evidence from the United States. J. Plan. Lit. **30**(4), 413–432 (2015)
50. S. Haghighatafshar, J. la C. Jansen, H. Aspegren, K. Jönsson, Conceptualization and schematization of mesoscale sustainable drainage systems: a full-scale study, Water **10**(8), 1–16 (2018)
51. J. N. Zouras, Collaborative decision-making in green and blue infrastructure projects: the case of copenhagen's hans tavsens park and korsgade KTH R (2020)
52. Á.-F. Morote Seguido, Educational proposal for the interpretation of the flood risk areas. Didáctica Geográfica **18**, 307–312 (2017)
53. P. Dircke, A. Molenaar, Climate change adaptation; innovative tools and strategies in Delta city Rotterdam. Water Pract. Technol. **10**(4), 674–680 (2015)
54. D. Tubridy, Green climate change adaptation and the politics of designing ecological infrastructures. Geoforum **113**(May), 133–145 (2020)
55. H. Pötz, T. Anholts, M. De Koning, Multi-level safety: water resilient urban and building design. Amersfoort (2014)
56. I. Roovers, Floating Amsterdam, the development of IJburg's Waterbuurt (2010), p. 19
57. E. English, Amphibious foundations and the buoyant foundation project: innovative strategies for flood-resilient housing. *Road Map Towards a Flood Resilient Urban Environment* (2009), pp. 25–27
58. E. English, N. Klink, S. Turner, Thriving with water: developments in amphibious architecture in North America. E3S Web Conf. (2016)
59. P. Nilubon, W. Veerbeek, C. Zevenbergen, Amphibious architecture and design: A catalyst of opportunistic adaptation?—case study Bangkok. Proc. Soc. Behav. Sci. **216**, 470–480 (2016)
60. G. Mihalakakou et al., Green roofs as a nature-based solution for improving urban sustainability: progress and perspectives. Renew. Sustain. Energy Rev. **180**, 113306 (2023)

61. F.R. Leite, M.L.P. Antunes, Green roof recent designs to runoff control: a review of building materials and plant species used in studies. Ecol. Eng. **189**, 106924 (2023)
62. C. Lamera, G. Becciu, M.C. Rulli, R. Rosso, Green roofs effects on the urban water cycle components. Proc. Eng. **70**, 988–997 (2014)
63. G. Battista, E. de Lieto Vollaro, P. Ocłoń, R. de Lieto Vollaro, Effects of urban heat island mitigation strategies in an urban square: a numerical modelling and experimental investigation. Energy Build. **282** (2023)
64. H. Liu et al., Impacts of green roofs on water, temperature, and air quality: a bibliometric review. Build. Environ. **196** (2021)
65. C. Loiola, W. Mary, L. Pimentel da Silva, Hydrological performance of modular-tray green roof systems for increasing the resilience of mega-cities to climate change. J. Hydrol. **573**, 1057–1066 (2019)
66. F. Rosso, S. Mannucci, M. Morganti, S. Mariani, M. Ferrero, The effect of sustainable urban drainage systems on outdoor comfort and runoff (2019)
67. L.M. Cook, T.A. Larsen, Towards a performance-based approach for multifunctional green roofs: an interdisciplinary review. Build. Environ. **188**, 107489 (2021)
68. A. Nardini, S. Andri, M. Crasso, Influence of substrate depth and vegetation type on temperature and water runoff mitigation by extensive green roofs: Shrubs versus herbaceous plants. Urban Ecosyst. **15**(3), 697–708 (2012)
69. Z. Zhang, C. Szota, T.D. Fletcher, N.S.G. Williams, J. Werdin, C. Farrell, Influence of plant composition and water use strategies on green roof stormwater retention. Sci. Total. Environ. **625**, 775–781 (2018)
70. C. Li, M. Liu, Y. Hu, T. Shi, X. Qu, M.T. Walter, Effects of urbanization on direct runoff characteristics in urban functional zones. Sci. Total. Environ. **643**, 301–311 (2018)
71. R. Meng, Q. Zhang, D. Li, H. Wang, Influence of substrate layer thickness and biochar on the green roof capacity to intercept rainfall and reduce pollution in runoff. Polish J. Environ. Stud. **30**(5), 4085–4103 (2021)
72. G. Zanin, L. Bortolini, Performance of three different native plant mixtures for extensive green roofs in a humid subtropical climate context. Water (Switzerland) **12**(12), 1–15 (2020)
73. M. Manso, J. Castro-gomes, Green wall systems: a review of their characteristics. Renew. Sustain. Energy Rev. **41**, 863–871 (2015)
74. Ö.T. Burhan, E. Karaca, Vertical Gardens. in *Advances in Landscape Architecture*, ed. by M. Özyavuz (2013)
75. S.A. Palermo, M. Turco, Green Wall systems: where do we stand? IOP Conf. Ser. Earth Environ. Sci. **410**(1) (2020)
76. G. Vox, I. Blanco, E. Schettini, Green façades to control wall surface temperature in buildings. Build. Environ. **129**, 154–166 (2018)
77. J. Coma, G. Pérez, A. de Gracia, S. Burés, M. Urrestarazu, L.F. Cabeza, Vertical greenery systems for energy savings in buildings: a comparative study between green walls and green facades. Build. Environ. **111**, 228–237 (2017)
78. V. Oquendo-Di Cosola, F. Olivieri, L. Ruiz-García, A systematic review of the impact of green walls on urban comfort: temperature reduction and noise attenuation, Renew. Sustain. Energy Rev. **162**, 112463 (2022)
79. F. Yan, J. Shen, W. Zhang, L. Ye, X. Lin, A review of the application of green walls in the acoustic field. Build. Acoust. **29**(2), 295–313 (2022)
80. Z. Azkorra et al., Evaluation of green walls as a passive acoustic insulation system for buildings. Appl. Acoust. **89**, 46–56 (2015)
81. J.T. Lau, D.Y.S. Mah, Green wall for retention of stormwater. Pertanika J. Sci. Technol. **26**(1), 283–298 (2018)
82. M. Turco, S.A. Palermo, M. Maiolo, B. Pirouz, P. Piro, Experimental and numerical analysis to assess the substrate hydraulic properties and the retention capacity of a green wall module. Urban Water J. **20**(1), 49–59 (2023)
83. M. Ostendorf, W. Retzlaff, K. Thompson, M. Woolbright, S. Morgan, S. Celik, Storm water runoff from green retaining wall systems. Cities Alive: Green Roofs and Green Walls Conference (2011), pp. 1–15

84. F. Ascione, R.F. De Masi, M. Mastellone, S. Ruggiero, G.P. Vanoli, Green walls, a critical review: knowledge gaps design parameters, thermal performances and multi-criteria design approaches. Energies **13**, 1–39 (2020)

Chapter 4
Navigating Futures: Scenario Planning in Urban Resilience and Climate Adaptation

Abstract The role of scenario planning in architectural and urban decision-making is critical for addressing the complexities and uncertainties inherent in the built environment, particularly in the context of climate change. By providing a structured framework to envision alternative futures, scenario planning facilitates strategic decision-making that is both robust and adaptable. This exploration delves into the theoretical underpinnings of scenario planning, analyzing its conceptual intricacies and applicability within urban planning. Emphasis is placed on integrating scenario planning with Decision Making Under Deep Uncertainty (DMDU) methodologies to equip professionals with the tools necessary for navigating future uncertainties. These methodologies enhance the capacity of planners to prepare for diverse future conditions, ensuring sustainable and resilient urban environments. This integrated approach is essential for fostering urban resilience and adaptability in the face of climate change. Through detailed analysis and a case study, the benefits and challenges of scenario planning and DMDU are illustrated, providing valuable insights for practitioners. By leveraging these advanced planning tools, urban planners can better navigate the dynamic and uncertain landscape of architectural and urban development, ensuring cities are prepared for future challenges and opportunities.

Keywords Scenario planning · Climate adaptation · Decision making under deep uncertainty (DMDU) · Adaptive planning · Exploratory scenarios

4.1 Framework for the Future: The Role of Scenario Planning in Architectural and Urban Decision-Making

In the preceding chapters, we explored the complexities and uncertainties inherent in the built environment, acknowledging the promising potential of scenario planning [1] as a strategic tool for planners in addressing these challenges [2]. Building upon this foundation, the present chapter extends the inquiry, comprehensively analyzing the literature surrounding scenario planning. This examination includes exploring

S. Mannucci, *Climate Adaptation in Urban Planning*,
SpringerBriefs in Architectural Design and Technology,
https://doi.org/10.1007/978-981-97-4106-9_4

both the potential benefits and inherent limitations associated with applying scenario planning in the context of the built environment.

Scenario planning, recognized for its forward-looking and dynamic approach, offers a structured framework for envisioning alternative futures. This chapter aims to elucidate the theoretical underpinnings of scenario planning, analyzing its conceptual intricacies and applicability.

Having framed the discourse on scenario planning, the discussion proceeds into the investigation of its application within Decision Making Under Deep Uncertainty (DMDU) [3]. The goal is to assess how these methodologies and tools can be effectively utilized by professionals and researchers in the fields of architecture and urban planning, integrating scenario planning and decision-making processes and offering insights into its tangible benefits and potential challenges within the specific context of the built environment.

This chapter aims to assist practitioners and researchers in enhancing the literature supporting adaptive planning in urban contexts. By integrating theoretical insights with practical considerations, the goal is to contribute to the ongoing discourse on strategic decision-making in the dynamic and uncertain landscape of architectural and urban planning amidst climate change. The ideas and concepts provided can be further explored to advance the real-world application of city adaptation.

4.1.1 Overview of Scenario Planning

Scenario planning is an essential strategic instrument for exploring future possibilities and has evolved significantly from its origins. It embodies a rigorous approach, providing a set of internally cohesive perspectives on possible futures without purporting to predict them with absolute certainty. This methodology facilitates a structured exploration of possible scenarios, enabling organizations and decision-makers to navigate uncertainty [4].

Over the past decade, the prevalence of scenario planning has substantially increased. Studies suggest a notable association between adopting scenario planning techniques and heightened uncertainty, unpredictability, and instability.

While scenarios can theoretically be crafted for any time frame, their utility tends to be more pronounced when developed for the long term. Leveraging scenario planning for long-range planning and strategic foresight proves instrumental in facilitating swift adaptations to significant changes [1].

What distinguishes scenarios is their divergence from predictions; they are not concerned with precise descriptions of the future but rather challenging existing paradigms and presenting a series of stories that direct attention to often-overlooked aspects [4].

Delving into the roots of scenario planning [5], in the RAND Corporation during the 1940s, Herman Kahn pioneered a technique called "*future-now*" thinking. Initially applied to research weapons technology, Kahn's approach combined detailed analyses with imagination, producing reports written from a future perspective. Kahn later founded the Hudson Institute, specializing in writing stories about the future to help people consider the "*unthinkable,*" such as the consequences of nuclear war [6, 7].

Around the same period, the Stanford Research Institute began offering long-range planning, considering political, economic, and research forces as primary business development drivers. This shift toward societal changes marked a turning point, shaping the trajectory of scenario planning.

Corporate sponsorship of the Hudson Institute by companies like Shell exposed the business world to this transformative line of thinking. Ted Newland of Shell played a pivotal role in encouraging the organization to consider future scenarios [8].

The success of Shell in navigating the Yom Kippur War and plummeting oil prices due to scenario planning encouraged other organizations to follow suit. However, the success was short-lived, as the 1980s brought recession and corporate staffing reductions, leading to a decline in scenario use.

The 1990s witnessed a resurgence in scenario planning, with corporations cautiously reintegrating its application. Once more embraced, the tool expanded its influence beyond individual companies to national levels, successfully bringing diverse groups together [5].

Over the years, scenario planning has evolved into distinct schools [9].

The Intuitive Logics School (Anglo-Saxon School), rooted in Herman Kahn's work [6, 7], involves identifying forces, ranking uncertainties, and constructing scenarios. Royal Dutch/Shell stands as a notable pioneer in this approach.

The Probabilistic Modified Trends School, influenced by methodologies like Trend Impact Analysis (TIA) and Cross-Impact Analysis (CIA) [10], focuses on generating multiple alternative futures, adjusting extrapolations based on historical data, and assessing interdependencies.

From Gaston Berger's emphasis on long-term planning, the French School (La Prospective) incorporates mathematical and systems analysis-oriented tools, combining various analytical methods for scenario creation with a focus on long-term planning [11].

Since the mid-twentieth century, scenario planning has gained popularity, witnessing diverse methods and techniques applied across the public and private sectors. It has been proven indispensable for addressing complex challenges, aiding strategic decision-making, navigating human reasoning limitations, recognizing future trends, and adapting to changes. It facilitates the identification of crucial factors contributing to uncertainty, fostering an understanding of their interplay and crafting plausible alternative narratives of the future [12]. Studies on scenario planning delve into its origins and typologies, emphasizing both qualitative and quantitative methodologies and techniques [1, 13–16].

Börjeson et al. [17] classified the typology of scenarios based on three key questions: "*What will happen?*", "*What can happen?*", and "*How can a specific target be reached?*". Predictive scenarios address the first question, aiming to predict future outcomes based on probability and likelihood. Explorative scenarios, tackling the second question, prove beneficial amidst uncertain conditions and rapid changes. Normative or prescriptive scenarios, addressing the third question, incorporate normative considerations and values to assess how to reach or avoid a specific target.

More in detail, predictive scenarios [18, 18] offer a detailed narrative of a specific future state, providing organizations with a comprehensive roadmap to anticipate challenges and adjust strategies based on existing trends, patterns, and available data. Particularly valuable in industries shaped significantly by historical data and trends, predictive scenarios allow organizations to paint a vivid picture of the future. Through analyses of existing data, they construct a narrative encompassing a wide range of possibilities and outcomes, facilitating more informed decision-making. Moreover, predictive scenarios serve as a valuable tool to test the viability of strategies in different future contexts. By simulating various scenarios, decision-makers can proactively evaluate their plans' resilience, identify potential gaps or weaknesses, and fine-tune strategies. Leveraging historical data and patterns, organizations can create scenarios that capture the nuances of their industry, guiding decision-making effectively.

Exploratory scenarios [18, 19] are employed to investigate a wide range of future possibilities, allowing for extensive exploration of different factors and uncertainties that may impact the organization. These open-ended scenarios offer valuable insights into risks and opportunities, providing a broad foundation for early-stage scenario planning. By delving into the intricacies of potential futures, planners gain an invaluable tool for comprehensively investigating factors and uncertainties influencing possible outcomes. These explorations help to provide assessments of risks and opportunities. Throughout the early stages of scenario planning, exploratory scenarios establish a solid base for further analysis and strategic choices, fostering a deep understanding of the multifarious paths the future may take.

Lastly, normative scenarios [20, 21] outline a preferred future state, guiding decision-making and shaping strategies toward specific desired outcomes.

The categorization of scenario types as normative occurs when the initiation of the scenario-building process involves a clearly defined objective or seeks participant inputs to identify targets initially. Normative scenarios typically involve the comparison of relative preferences among these targets or an effort to determine strategies for achieving them.

4.1.2 Scenario Planning for the Built Environment

In the latter part of the twentieth century, the waning impact of ambitious urban schemes led to critiques of contemporary planning for its departure from utopian ideals. This transition is characterized by a shift from grand visionary projects to more pragmatic, single-dimensional projections. This transformation was partly a reaction to the unfulfilled promises of comprehensive plans that frequently culminated in unwelcoming spaces, thereby undermining the credibility of utopian thought within the planning discourse [22].

As discussed in Chap. 2, the 1960s witnessed a significant transformation in urban planning, shifting focus from utopian projects and physical city design to emphasizing quantitative techniques and rational processes. The applied science of planning sought optimal actions and policies to achieve identified goals, employing forecasting techniques and expert design to simulate system functions and select optimal plans from a range of alternatives. However, the rational planning approach faced criticism as planning was increasingly recognized as a political process, leading to new methodologies inspired by business and strategic planning [23].

While connecting broader resources and goals to specific objectives, the advent of strategic planning lacked a spatial dimension. This limitation spurred a theoretical shift toward *communicative rationality* [24]. With the ascendancy of participatory approaches, rationality ceased to be solely logic-based, evolving into an informed consensus fostered by a community of individuals in a specific place and time [2, 25]. Stakeholder participation and scenario development, rooted in participatory methodologies, paved the way for the transition from expert-led blueprint planning to a more dynamic, process-oriented, and collaborative planning paradigm [23] (see Chap. 2, Fig. 2.5).

Traditionally, planners concerned with the design of the built environment have often leaned toward selecting a single preferred future from a limited set of alternatives. However, the recognized idea that the built environment is complex and subject to endogenous and exogenous uncertainties fostered the shift toward new approaches [2]. Furthermore, the advancements in computer tools and the growing consensus that the traditional approach of attempting to predict the future is futile shifted the focus toward the adoption of scenario planning [26–28].

The scenario planning paradigm rejects the notion of good or bad scenarios, presenting a collection of plausible alternatives, some of which may introduce new and occasionally uncomfortable possibilities [26].

Initially inspired by methodologies derived from corporate strategic planning, scenario planning advocates for the creation of multiple plausible scenarios rather than relying on a single preferred vision or a forecast as the foundation for planning [19, 26, 29].

Specifically, urban planning and design conventionally relied on scenario planning techniques to define a preferred or normative scenario. However, a recent shift among practitioners moves beyond normative scenarios, embracing the exploration of a broader spectrum of uncertain futures, termed exploratory scenario planning. The normative tradition in scenario planning centers on crafting a scenario that depicts a desired future, typically accompanied by scenarios illustrating competing visions or futures resulting from ongoing trends. However, normative approaches often neglect uncertainties about the future [19].

In contrast, recent developments in practice involve a different approach to scenario usage, creating scenarios that depict possible futures independently of whether stakeholders view them as desirable. The objective of employing scenarios in this manner is to explore and prepare for uncertainties such as economic growth variations, climate change impacts, and the introduction of new technologies [30].

The focus is on making strategic decisions in the present that align with plausible futures. The scenario-building process prioritizes effectiveness over accuracy, aiming to facilitate better decisions in the future [31]. Practicing responses to anticipated events enhances planners' preparedness through heightened sensitivity to the planning environment and accelerating response rates. It is a proactive approach to uncertainty management [32].

As Klosterman discussed [27], when dealing with multifaceted and intricate planning situations, urban planners necessitate a typology of scenarios that transcends narrow considerations and aligns more suitably with complex system planning. Exploratory scenarios, in particular, represent a departure from the traditional predictive models, pivotal in navigating uncertainties and envisioning alternative futures, facilitating the introduction of a more flexible and adaptive approach to planning [33].

The prevalent methodology employed for the development of exploratory scenarios in urban planning derives its roots from a process that broadly involves a four-step process [19, 29, 34, 35]. First, there is a phase of brainstorming and discussion to identify key trends, constraints, and issues. Next, driving forces are identified, categorized as assumptions or uncertainties, and evaluated based on their degree of uncertainty and impact. Subsequently, the most uncertain and impactful driving forces are selected as the foundation for building scenarios. The final step involves constructing the narrative based on these driving forces [36]. Although various exploratory scenario-generating processes yield similar summaries on how uncertain driving forces shape scenarios, they often undergo further development by incorporating quantitative impact models.

Indeed, an essential aspect of current practices involves identifying critical uncertainties. These uncertainties are assessed to gauge their potential impact on urban development and their likelihood and magnitude of influence, as rating uncertainties add layers to the scenario-building process and facilitate the decision-making regarding the possible planning strategies [37].

Following the analyses of uncertainties, scenario development adopts a narrative form, presenting qualitative stories that unfold in response to different combinations of these uncertainties. This narrative approach allows for a holistic exploration of potential futures, considering the interplay of various factors. Visualization tools ranging from immersive storytelling techniques to digital simulations are employed to enhance the communicative power of these scenarios [30]. With technological advancements, there is a growing trend toward leveraging data analytics and machine learning to bolster the robustness of scenario development. These tools hold the potential to quantify uncertainties, model complex interactions, and derive data-driven insights, thereby providing a more evidence-based foundation for decision-making in urban planning.

Insights drawn from the practical application of exploratory scenarios in urban projects offer valuable lessons for practitioners. Rather than prescribing a rigid framework, these cases suggest considerations that practitioners may find beneficial as they navigate their own scenario-planning initiatives. Scenario performance is then evaluated using a diverse set of metrics or indicators. Another significant consideration is identifying robust and contingent strategies and projects for inclusion in plans. Robust strategies perform well across all scenarios, while contingent strategies may be context-specific [38]. Advanced projects could even outline events or conditions acting as triggers for implementing particular ideas, a practice observed in water planning [39].

Nevertheless, challenges persist in urban scenario planning applications, ranging from insufficiently addressing uncertainties to overemphasizing selecting a preferred future and a deficiency in effective public involvement [38, 40]. It is also difficult to translate exploratory scenarios into concrete plans and actions.

Exploratory scenarios, by their nature, describe a range of possible futures, including both desirable and undesirable outcomes. This can make it challenging for planners to develop specific recommendations that are robust across all scenarios.

Furthermore, the use of complex modeling and data analysis in scenario development can be time-consuming and resource-intensive, requiring expertise and technical support. Despite the insights offered by exploratory scenarios into potential trajectories, challenges persist in translating them into actionable strategies. There is a recognized need for clearer pathways from scenario development to concrete urban planning decisions. The integration of these scenarios with traditional planning tools and regulatory frameworks remains an ongoing challenge, as does ensuring that the narratives generated resonate with the broader community [22].

Lastly, a further challenge within the domain of exploratory scenarios lies in the absence of a comprehensive framework that unequivocally delineates the systematic approach for addressing uncertainties and effectively incorporating the scenario narrative into the design of an actionable plan [19]. Despite the increasing adoption of exploratory scenarios in urban planning, a notable deficiency exists in methodologies that systematically tackle uncertainties and integrate scenario narratives into pragmatic planning endeavors. This impedes translating insights from exploratory scenarios into practical, well-informed decisions.

In order to address the challenges above, the subsequent section will introduce the Decision Making under Deep Uncertainty (DMDU) methods and tools. This ensemble of approaches has demonstrated remarkable efficacy in assisting decision-makers and planners in navigating deep uncertainty and formulating plans that are not only robust but also adaptive.

4.2 Decision Making Under Deep Uncertainty

Decision-makers grapple with a multitude of uncertain factors that have the potential to impede the success of their plans. The ongoing dialogue between policy stakeholders and scientists from diverse fields is essential for effectively communicating uncertainties [41]. While urban planners commonly rely on forecasts, the risk of undesired outcomes looms if the future deviates from the assumptions embedded in these forecasts.

As introduced in Chap. 2, uncertainty, a multi-dimensional concept, is defined as any deviation from the unattainable ideal of determinism [42]. Specifically, deep uncertainties are particularly significant, as no forecasting tool can provide aid to deal with such events. They constitute a distinct level of uncertainties characterized by the inherent challenge of systematically enumerating alternative assumptions and assigning them a ranked order based on plausibility. Unlike uncertainties of a more straightforward nature (levels one and two), wherein various potential outcomes can be delineated and assessed for their likelihood, deep uncertainties lack a clear consensus or predictability regarding the relative probabilities of different scenarios. These uncertainties are emblematic of intricate, multifaceted factors that defy straightforward quantification or definitive ranking [43].

Decision Making under Deep Uncertainty (DMDU) has emerged as a comprehensive umbrella encompassing diverse planning and decision-making approaches aimed at crafting robust and adaptive plans [44–46]. Various approaches offer different analytical perspectives and techniques [47–50], and their combination can prove beneficial in specific contexts [44–46, 51].

Forecasting methods can support the first two levels of uncertainty but have been proven to contain "*fundamental weaknesses.*" [52] Therefore, DMDU cannot be based just on a prediction about the future, which implies a change from the paradigm "*predict-then-act*" to "*monitor and adapt*" [53].

When grappling with complex systems undergoing dynamic changes due to deep uncertainties, understanding the system's functionality and anticipating future outcomes becomes challenging. The utility of simulation models becomes evident in supporting decision-making under deep uncertainty, particularly when dealing with complex dynamic systems [54, 55]. Simulation models effectively instantiate the system of interest, allowing for exploring plausible futures and testing interventions in light of the inherent uncertainties [56–58].

Several methodologies and tools prove invaluable in supporting Decision Making under Deep Uncertainty (DMDU) when dealing with complex and dynamic systems. Kwakkel and Haasnoot [51] presented a taxonomy categorizing available approaches and highlighted key concepts within the DMDU paradigm:

Exploratory Modeling: This involves a computer-assisted research methodology to explore complex systems that surpass full comprehension through human reasoning alone.

Decision Support: Encompassing approaches and tools intended to facilitate understanding of a problem and the potential course of action, often implemented through a posteriori decision analysis.

Adaptive Planning: This approach enables the design of plans that can adapt as conditions evolve. It entails exploring a broad spectrum of possible futures to construct a plan with flexible, adaptive actions capable of responding to potential uncertainties that may threaten the plan's success.

4.2.1 *Generic Elements for Application*

Under the umbrella of DMDU exists a family of approaches, each with a specific focus. A generic framework to support policy problems can be used to categorize these approaches and make an informed selection for application in a specific case.

Therefore, to structure the problem and organize the available information, the "XLRM" framework [44, 59] is used, and the elements for the analysis are divided into four groups:

- Exogenous uncertainties ("X"): Elements that planners cannot control but could affect the feasibility or success of the decided strategies. The uncertainties can be environmental, demographical, economic, etc.
- Policy levers ("L"): Short-term strategies applied to reach near-term objectives.
- Relationships ("R"): An essential connecting element that delineates interactions among the specified groups throughout time is typically embodied by a model aligning with the planners' objectives.
- Measures ("M"): Elements used to evaluate the performance of the policy levers in the different scenarios generated. They are ranked according to the desirability of the outcome, according to the planners' aims.

Following the organization of information, the iterative utilization of the simulation model is employed to assess proposed actions across numerous plausible futures and ascertain the most resilient strategy. In simpler terms, this refers to the strategy exhibiting better performance or lesser susceptibility to uncertainties. Given uncertainties, the literature underscores the significance of robustness in planned actions. Consequently, DMDU approaches have emerged to facilitate the development of robust plans. Various approaches offer distinct analytical perspectives, and combining or interchanging techniques can be advantageous depending on the specific context [51].

As summarized by Vincent A.W.J. Marchau et al. [60], each DMDU here addressed comprises the following elements and steps:

1. Frame the Analysis

Formulate triggering issue (problem or opportunity): This initial step establishes the context and sets the boundaries for the entire analysis. It involves examining the questions or issues at hand, clarifying constraints, identifying stakeholders, and determining the analytical approach to be followed.

Specify system and its boundaries: This step defines the system of interest and its boundaries, crucial for subsequent analysis. The system model, designed and utilized in later steps, serves as an exploration tool rather than a predictive tool.

Specify objectives/goals and outcome indicators: Identifying policy objectives or goals and related outcome indicators is essential in this step, focusing on making informed choices among alternative policies.

Identify alternative policies: This step involves specifying policies for assessment using the system model, aiming to understand and refine the nature of policies, discover additional ones, or modify and combine existing ones.

2. Perform Exploratory Uncertainty Analysis

Delineate uncertainties or discrepancies related to external forces, system structure, outcome indicators, and the valuation of outcomes: Pinpoint key areas of uncertainty influencing policy decisions, encompassing external forces, system responsiveness, system outcomes, and the significance of outcomes to stakeholders.

Examine the consequences of the policies and weaknesses: Conduct exploratory analysis, frequently employing techniques like Exploratory Modeling, to investigate an extensive array of assumptions and scenarios. This includes scrutinizing how different policies fare under diverse conditions and identifying factors contributing to their success or failure.

3. Choose Initial Actions and Contingent Actions

Illustrate tradeoffs detected by analysis and means for future adjustments: Tools such as Scenario Discovery inform further iterations of the model, exploring potential approaches to addressing uncertainty through policy design changes.

Select and plan for the adoption of an initial policy and mechanisms for future adjustments: Based on earlier results, initial actions are chosen to perform well under uncertainties, and plans for future contingent actions are prepared to respond to uncertain events.

Plan for monitoring and making possible adaptations to the policy: In addition to defining the (initial) policy, a "signpost monitoring system" is outlined to monitor the validity of assumptions, the progress of implementation, and the need for timely and effective policy adjustments.

4.2.2 Key Elements and Main Approaches

This section introduces a comprehensive exploration of the key elements previously introduced, aimed at providing a nuanced and in-depth understanding of the DMDU methods and tools.

The key elements mentioned in Kwakkel and Haasnoot's taxonomy, as previously mentioned [51] are Exploratory Modeling [61], Decision Support [62], and Adaptive Planning [63].

Exploratory Modeling [57] constitutes a computational, experimental research method delving into various "*what if*" scenarios within a complex system under conditions of deep uncertainty. The objective is not to create a singular model replicating reality, given that a single representation is an unreliable source of information for the targeted outcomes [64]. Instead, the approach explores the system's behavior across diverse alternative assumptions regarding uncertainties. Consequently, models are generated to analyze the system's behavior amidst uncertainties and discern the conditions negatively impacting the plan's success. Numerous computational experiments are conducted, and their results are scrutinized to identify relevant scenarios and uncertainties pertaining to the problem. Exploratory Modeling aids in pinpointing a pertinent subset of models through search techniques, thereby assisting planners in assessing the consequences of uncertainties on the system and the outcomes of interest.

The process of identifying relevant subsets of models is commonly termed vulnerability analysis. The fundamental idea is to identify key subspaces in the uncertainty space where candidate strategies fail to meet specified objectives. The same analysis technique can be applied to the lever space to identify subspaces that perform either well or poorly. Vulnerability analysis explores uncertainties and policy levers using two complementary analysis techniques: global sensitivity analysis [65] and subspace partitioning. Global sensitivity analysis helps reduce the problem's dimensionality by pinpointing uncertainties and/or policy levers significantly influencing the outcomes of interest. Subspace partitioning, also known as Scenario Discovery [66, 67], focuses on identifying a subspace in the uncertainties and/or policy levers space generating (un)desirable outcomes. Statistical or data mining algorithms are utilized in subspace partitioning to find subspaces in a multi-dimensional input variable space of uncertain model inputs that produce noteworthy outcomes in the output.

The exploration of uncertainties and policy levers can be accomplished through both "exploration" and "search." Exploration involves analyzing the spaces of uncertainties and policy levers, providing insights into how they affect the outcomes of interest, and identifying circumstances where planners can achieve objectives. Conversely, search strategies aim to identify specific locations within the uncertainties or policy levers spaces with desired properties, such as best or worst possible outcomes.

Another approach involves pre-specifying uncertainties or policy levers, obviating the need to explore the entire space of policy levers or uncertainties. Instead, the focus shifts to assessing how pre-specified levers' performances are impacted by uncertainties or evaluating the effects of uncertainties if the assumptions made were accurate. Leveraging Exploratory Modeling, policy levers can be stress-tested over numerous plausible futures to assess the robustness of selected strategies to unforeseen events.

A pivotal concept in the DMDU literature is robustness, where DMDU methods aim to assist planners and decision-makers in identifying solutions that perform well across a broad spectrum of future scenarios. Various robustness metrics, including satisficing and regret metrics, exist in the literature [68]. Regret metrics trace their origins to the min–max regret approach [69], comparing the chosen policy's performance in a scenario with the best option for that same scenario. The goal is to identify solutions with low regret across all scenarios. Satisficing metrics, inspired by Simon [70], seek policies capable of meeting satisfactory performance in as many scenarios as possible, often categorized based on a performance threshold. The objective is to maximize scenarios where a candidate policy meets the threshold. While DMDU approaches employ diverse tools and strategies, common components underpin all the approaches.

Decision Support [51, 71] aids in addressing complex and uncertain systems, particularly in scenarios involving multiple decision-makers seeking consensus. The decision-making process in such contexts necessitates an iterative approach that fosters learning across diverse problem framings, stakeholder preferences, and tradeoffs.

In the iterative approach inherent to decision support under deep uncertainty, candidate policy decisions are introduced into the analysis through stress testing across a broad spectrum of uncertainties. The subsequent characterization of uncertainties focuses on their impact on decision outcomes.

This conceptualization of decision support transcends the traditional a priori agreement on assumptions about the probability of alternative states of the world and the aggregation of competing objectives. Instead, it embraces an a posteriori exploration of tradeoffs among objectives and the robustness of performance across potential futures. Decision support shifts from attempting to prescribe the "right" choice and pivots toward enabling deliberation and collaborative sense-making among diverse decision-making entities.

Adaptive planning [45, 72, 73] entails intentionally designing plans with inherent flexibility, allowing for continuous adjustments over time in response to unfolding future conditions. Unlike traditional approaches, where adjustments are made reactively after the fact, adaptive planning involves announcing the plan's adaptability simultaneously with its formulation.

The core objective of adaptive plans is to achieve decision robustness by acknowledging and accommodating the resolution of deep uncertainties as the future unfolds. Decision-makers can dynamically adapt the plan's implementation based on emerging information, contributing to a more resilient decision-making process. To facilitate such adaptability, extensive exploration of a diverse array of potential futures is imperative during the initial design of the plan.

The effectiveness of adaptive planning hinges on acquiring insights into which actions align best with specific future scenarios. Additionally, decision-makers must identify key signals from the evolving future that warrant close monitoring, ensuring the timely implementation of appropriate actions. Unlike planned adaptation, where changes occur at predetermined intervals, adaptive planning involves a paradigm shift. It focuses on planning that is conditional on observed developments rather than adhering to a fixed timeline. Consequently, the timing of plan adaptation remains uncertain, contingent on the unfolding dynamics of the future. This distinction underscores the dynamic and proactive nature of adaptive planning in navigating complex and uncertain decision landscapes.

Under the aegis of DMDU, a multitude of approaches and tools aim to assist decision-makers in navigating complex and uncertain landscapes. For the purposes of this book and to provide clarity to readers approaching DMDU for the first time, two approaches are introduced that can be particularly interesting for applications within the built environment. These approaches, namely Robust Decision-making (RDM) and Dynamic Adaptive Policy Pathways (DAPP), offer distinct yet complementary perspectives on addressing uncertainties in decision-making.

As defined in the literature, the RDM [66] framework serves as a versatile approach for decision-makers grappling with deep uncertainty. Developed primarily by researchers at the U.S.-based RAND Corporation, RDM integrates Decision Analysis, Assumption-Based Planning, Scenario Analysis, and Exploratory Modeling. It offers a structured, iterative process encompassing five key steps [74]. The initial steps involve generating and framing decisions, strategies, or alternatives based on stakeholder preferences and key factors, followed by the generation of different scenarios. Performance metrics are outlined, and decisions are evaluated across scenarios. The ensuing steps involve isolating successful decisions, performing vulnerability and tradeoff analyses, and iteratively adapting alternatives to enhance robustness over a multitude of plausible future states. An illustrative example is the long-term water resources planning for the Colorado River Basin [75], where portfolios of strategies were evaluated across numerous scenarios, highlighting tradeoffs and vulnerabilities.

Complementary to RDM, the Dynamic Adaptive Policy Pathways (DAPP) [76, 77] framework emphasizes explicit decision-making over time. This approach allows decision-makers to assess sequences of decisions (pathways) over various futures, choosing the most promising alternative while maintaining flexibility for future decisions. DAPP integrates Adaptive Policymaking [78] and Adaptation Pathways [63, 79] into a unified framework. Developed primarily by researchers at Delft University of Technology, Deltares, and the World Bank, DAPP enables the evaluation of pathways and scenarios, incorporating adaptation tipping points and continuous monitoring. DAPP proves effective in infrastructure planning, as demonstrated in the case study of Amsterdam Airport Schiphol, showcasing its adaptability and resilience under diverse uncertainties.

As discussed by Kwakkel et al. [49], both RDM and DAPP can be used to design effective, flexible strategies, and it is even possible to combine them according to context-specific necessities. RDM and DAPP exemplify adaptability and resilience in decision-making under deep uncertainty, with RDM emphasizing iterative evaluations across diverse scenarios and DAPP emphasizing explicit decision-making over time, incorporating adaptation tipping points and continuous monitoring. Despite their demonstrated effectiveness, ongoing research and development are essential to refine and broaden the applicability of these approaches across diverse contexts.

4.3 Applicative Example for the Built Environment

Despite the growing attention toward scenario planning for urban contexts and the cutting-edge approaches and methods developed under the umbrella of DMDU, the literature combining the two approaches is still in its infancy. In this section is provided an example that is described in [80].

Broadly speaking, the case analyzed is structured according to the following steps:

Case Study Selection: The initial phase involves the identification of a complex system operating within uncertain conditions.

Data Collection: The data collection phase aims to comprehensively understand the chosen complex system's current state and potential dynamics.

Model Implementation: Through the collected data, a model is developed to simulate the interactions and behaviors of the analyzed factors within the complex system.

Information Organization: This step involves structuring and organizing the acquired information using the XLRM framework.

Exploration, Clustering, and Selection of Decision-Relevant Clusters: Various scenarios are generated to explore potential outcomes and responses within the complex system. The ensemble of experiments is created using a space-filling experimental design over the entire space spanned by a set of uncertain factors.

Scenario Discovery and Scenario Narratives: Scenario Discovery [66, 81] is applied to the identified relevant clusters with decision-making significance to analyze the experiment results. The outcomes of Scenario Discovery lay the groundwork for constructing scenario narratives, offering holistic depictions of the plausible evolution of the complex system under distinct conditions.

Further details to follow these steps are provided below. Notwithstanding, a full explanation and an in-depth analysis can be found in [80].

For the applicative example, the selected study area comprises an urbanized flood-prone deltaic region within Rome's X Municipality, situated between the Tiber River and the Tyrrhenian Sea [82]. Municipio X, situated in Rome, is comprised of distinct subdivisions, each characterized by its unique urban fabric (Fig. 4.1).

This area is of particular interest due to its existing vulnerability to flooding, which is expected to be exacerbated by future urban development and climate change.

Municipio X is divided into subdivisions, each characterized by a distinct urban fabric (see Fig. 4.1). Despite being densely populated, with a population density of 1533.9 inhabitants per square kilometer, Municipio X maintains a relatively low density compared to other districts, owing to several natural reserves within its borders.

The district's development is closely intertwined with Rome's broader urban dynamics, influenced by a combination of planning actions and socioeconomic factors. Informal settlements and unauthorized constructions emerged in postwar period due to housing challenges and socioeconomic disparities in Rome at the time.

In brief, the widely studied proliferation of informal settlements in Rome [83–87] was fueled by exponential population migration toward Rome, exacerbated by inadequate public housing policies that failed to provide viable alternatives to informal constructions.

Specifically, during the 1950s, as Rome's economy boomed and agriculture diminished in importance, many areas with a main agricultural vocation became hubs for informal housing to meet the rising demand. Consequently, an informal city emerged, albeit one now affected by flooding due to inadequate drainage systems designed for assumed agricultural land use.

Efforts to regularize informal settlements through building amnesties in 1985, 1994, and 2003 aimed to address the issue connected to the lack of planning interventions to cope with the housing demand. However, these actions only provided limited relief, failing to resolve the underlying problems associated with illegal constructions [84]. Today, regulatory and planning mechanisms continue to struggle

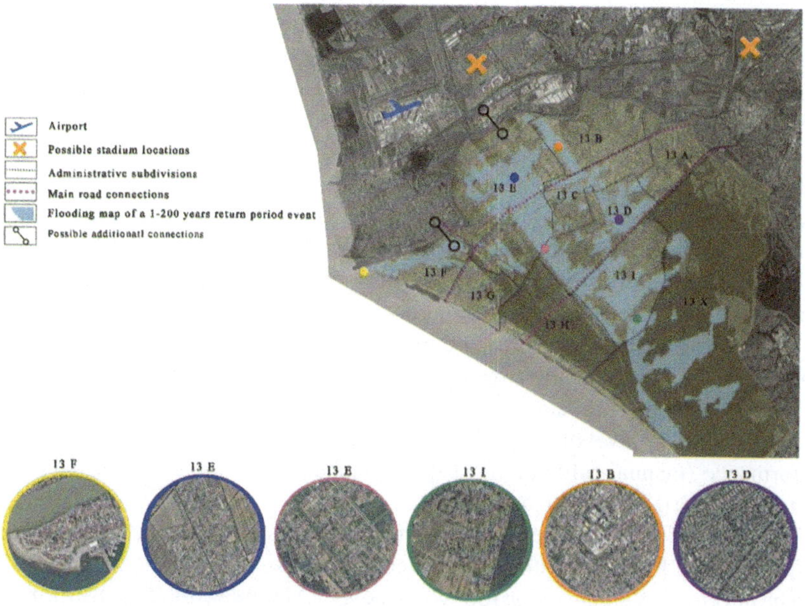

Fig. 4.1 Muncipio X of Rome and its subdivisions. In the image are highlighted the main road connections, the possible location suitable for the Stadium implementation, the adjacent international airport, and two possible additional connections, specifically a new bridge in 13 E and the enhancement of the bridge connecting 13 F with the other side of the river. In the light blue area is superimposed on the flooded area in a 1-200-year return period event. The circles present the different urban tissues more exposed to flooding. 13 F is the only recognized informal settlement. 13 E (pink), 13 I, and 13 B are recognized as former informal settlements awaiting regenerative intervention. The remaining areas 13 E (blue) and 13 D are tissue derived from planned interventions

to meet housing needs and manage land use effectively, perpetuating the conditions that initially led to uncontrolled expansion.

The case study is particularly interesting for the purpose of the investigation as it exhibits compelling features to explore, such as the urban development influenced by both formal and informal settlements, with former informal settlements posing higher vulnerability to climate change impacts. Furthermore, the administration was considering various infrastructure investments that could significantly alter the area's future urban landscape. Namely, the enlargement of a bridge in subdivision 13 F, a new bridge in 13 E and the possible construction of a football stadium in two different locations (see Fig. 4.1).

Therefore, to investigate the possible future development in the area, a land use change model based on historical data resulting from mapping the historical development was implemented.

A Geographical Information System (GIS) with a level III classification in the National Land Cover Map is used for the historical mapping.

Specifically, the data shows that the urban expansion in Municipio X has exhibited a consistent upward trend, primarily fueled by the expansion of "buildings" and "impervious non-built area" land use categories. More in-depth, the growth in "buildings" from 1949 to 1962 amounted to 206%; from 1962 to 1988, it stood at 172%; and from 1988 to 2017, it reached 53%, resulting in an overall increase of 1182% from 1949 to 2017. Similarly, the expansion in the "impervious non-built area" category saw growth of 344% from 1949 to 1962, 187% from 1962 to 1988, and 95% from 1988 to 2017, culminating in a total increase of 2381% from 1949 to 2017. The notable surge observed from 1949 to 1962 aligns with the post-war economic boom in Italy, which concurrently attracted migration toward Rome.

To analyze the potential impacts of future urban developments resulting from potential new infrastructure projects and/or new centers of attraction in the study area, we utilized the Metronamica land-use model [88]. Calibration and validation of the model were conducted using two distinct time periods: 1962–1988 for calibration and 1988–2017 for validation. The parameters for the calibration were adjusted until a satisfactory level of similarity was achieved, indicating the model's adequate performance (manual calibration) [89]. Subsequently, we simulated developments from 1988 to 2017 for validation purposes.

The accuracy of the model was quantitatively assessed using the Kappa statistic [90] and Fuzzy Kappa [91] through map comparisons.

The information organization following the XLRM framework (Fig. 4.2) primarily revolves around the demand for various land use classes and the impact of new infrastructures and attractions (X) (refer to Table 4.1). The range of values selected is derived from analyses of urban development, utilizing data extrapolation from historical demand patterns for each land use class Instead, as concerns the infrastructures' weight, the relative significance ranges from 0 to 1. The parameter ranges are derived from the values employed during the model calibration to determine the impact of accessibility parameters on the land use classes. Drawing upon past data, the study simulates potential outcomes under similar conditions through the land use model (R), considering the implications of climate change on the built environment. Notably, this case study excludes explicit policy interventions. The outcomes comprise 5000 land use maps for the year 2050, generated during the exploratory phase.

The computational experimentations that generated the 5000 maps are performed using the Modelling and Analysis Workbench (EMA) [57, 92]. The EMA workbench facilitates an exploration of the urban evolution in the X Municipality until 2050,

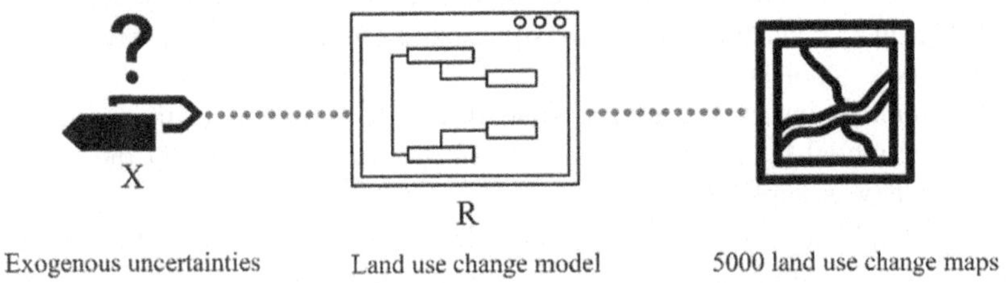

Fig. 4.2 Organization according to the XLRM framework for the case study

Table 4.1 Uncertainties considered and the associated rages of values

	Range
Land demand	
Informal settlements land use class	208–7051
Residential land use class	38,500–52,000
Residential + commercial land use class	4500–5500
Commercial land use class	1200–2300
Infrastructure weight	
Bridge subdivision 13 F	[0, 0.1, 0.4]
Bridge subdivision 13 E	[0, 0.1, 0.4]
Stadium (both locations)	[0, 0.1, 0.4]

generates, executes, and analyzes computational experiments structuring information using the XLRM framework. To explore the 8-dimensional uncertainty space, 5000 computational experiments are generated using Latin Hyper Cube Sampling [93].

Before Scenario Discovery [66, 81], the resulting land use maps from the experiments are clustered based on spatial land-use patterns. Experiment results are clustered, acknowledging the limitations of binary classification in complex models with rich outputs [94–97]. Clustering entails dividing a set of objects into subsets with shared characteristics, with each subset exhibiting unique attributes. Determining the optimal clustering results is crucial, as various algorithms can be employed for this purpose.

Two clustering algorithms, agglomerative clustering and K-medoids [98–100], were compared to identify the most suitable algorithm for the dataset. For this analysis, the kappa metric is utilized to specify the similarity between data points, which is consistent with model calibration and validation practices.

Both clustering algorithms determine the appropriate number of clusters based on within-cluster similarity, between-cluster dissimilarity, and interpretability. Agglomerative clustering emerges slightly superior, resulting in five clusters.

Once the scenarios are generated and clustered, the subsequent critical phase involves identifying and selecting clusters that bear particular significance for decision-making. This stage recognizes the significance of domain-specific knowledge and expertise held by the decision-maker. Leveraging their profound comprehension of the complex system, the decision-maker utilizes their expertise to identify clusters with the highest potential relevance and impact.

Visual inspection and quantitative analysis of each cluster's maps are conducted to identify decision-relevant clusters. In this case, the process involved quantifying the growth of each land use class and comparing results with 2017 maps to assess percentage increases. These evaluations are tailored to the case context, which is particularly relevant given the area's history of informal urbanization, heightening the vulnerability to climate-related phenomena and social disparities.

Clusters 4 and 5 represent the most undesirable outcomes (Fig. 4.3), exhibiting significant increases in informal settlements. To analyze the two selected clusters,

Scenario Discovery [73, 92] was employed on the identified clusters deemed significant for decision-making purposes. The findings of Scenario Discovery form the basis for crafting scenario narratives, which offer comprehensive depictions of the potential evolution of the intricate system under varied conditions. Each narrative provides a detailed portrayal of the interactions among variables. These scenario narratives are a resourceful tool for planners, offering a nuanced comprehension of the system's behavior across different future scenarios and their associated uncertainties. With this knowledge, planners can make well-informed decisions, develop adaptive strategies, and formulate contingency plans that align with the system's inherent dynamics. By translating complex analytical outcomes into understandable narratives, this process enhances communication between analysts and stakeholders.

Based on the findings of Scenario Discovery and context-specific information, two scenarios have been structured to address the needs of decision-makers and necessitate adaptive planning measures to mitigate undesirable outcomes:

Scenario 1—In Cluster 4, the land use pattern observed is influenced by the construction of a football stadium across the river, driving significant development in the built environment. Residential land-use demand experiences a substantial increase of 2.07% from 2017 to 2050, indicating Municipio X's appeal and the emergence of new job opportunities. However, this scenario also highlights heightened flood risk due to increased imperviousness, particularly in vulnerable zones initially developed informally or during building speculation. Climate change exacerbates urban heat island effects and precipitation, elevating pluvial flood risk. The rise in informal settlements by 0.19% underscores shortcomings in addressing housing demand, especially for lower-income residents, indicating the inadequacy of planned developments or the need for social housing initiatives.

Scenario 2—In Cluster 5, the scenario unfolds if the stadium is built across the reviver and the new attraction pole holds significant importance. Although residential land-use demand falls within the middle range, urban growth from 2017 to 2050 registers the lowest increase of 1.05% among clusters. This scenario underscores the high significance of the new attraction pole but reveals a mismatch between housing demand and development, resulting in a surge of informal settlements by 0.4%. Historical trends resurface, driven by increased attraction and housing demand, yet inadequate residential development exacerbates housing shortages and informal settlement growth. Vulnerabilities include the concentration of new developments in low-lying areas, widening flood zones, and heightened risk from extreme rainfall events, portraying a precarious situation concerning informal settlements.

4.3.1 Limitations

The built environment, characterized by its intricate web of interconnected factors, presents a myriad of uncertainties that profoundly influence urban planning

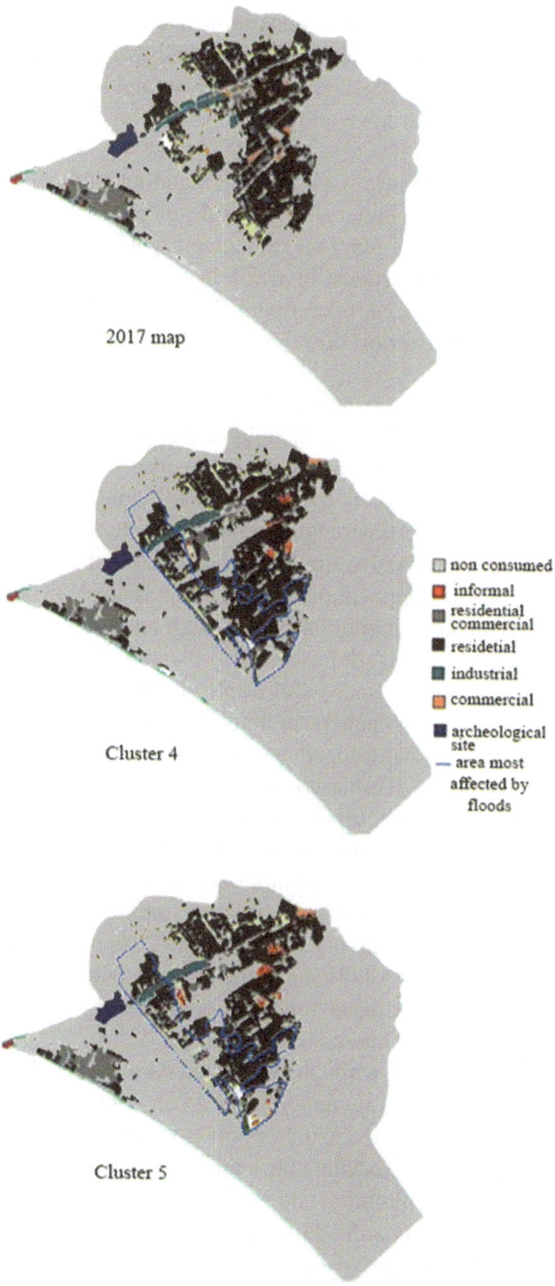

Fig. 4.3 The 2017 map and the selected clusters represent the most undesirable growth outcomes, specifically in the area highlighted, which is particularly vulnerable to floods. Author elaboration from [80]

endeavors [101]. These uncertainties stem from various sources, including demographic shifts, economic fluctuations, technological advancements, and environmental changes. As urban areas evolve and become increasingly complex, planners are tasked with navigating this intricate landscape to anticipate and address future challenges.

Despite advancements in urban planning methodologies over the past two decades, there remains a noticeable gap in developing a systematic framework to guide planners in crafting adaptive strategies capable of flexibly responding to uncertainty. While the importance of flexibility and adaptability in urban planning has been widely recognized, operationalizing these concepts into actionable plans requires a comprehensive workflow that accounts for uncertainties at every stage of the planning process.

The provided example introduces a generic workflow designed to assist urban planners in developing flexible adaptation plans amidst uncertainty. In the context of the case study presented, the workflow offers valuable insights for planners and decision-makers by systematically analyzing uncertainties related to land-use demands and infrastructure impacts, planners can better anticipate future challenges and devise proactive strategies to mitigate risks. However, it is essential to acknowledge that the application of DMDU approaches in the built environment is still in its infancy, necessitating further research and refinement.

While computational scenario planning holds promise as a tool for addressing uncertainties, its successful implementation often requires interdisciplinary collaboration. Urban planners, architects, engineers, environmental scientists, and policymakers must work together to develop holistic solutions that account for the multifaceted nature of urban challenges. Planners can effectively leverage diverse expertise to tackle complex urban issues by fostering collaboration across disciplines.

Despite the potential of computational modeling and scenario planning, it is crucial to recognize the limitations of this approach. Narrowing the focus solely on land-use demands and infrastructure impacts may overlook critical social parameters essential for addressing broader urban challenges, such as social equity and community resilience. Additionally, the absence of a hydraulic model in flood risk analysis underscores the need for more comprehensive modeling approaches considering urban systems' interconnected nature.

Furthermore, while the presented case study offers valuable insights, its findings may not universally apply to all urban contexts. Each city's unique characteristics and challenges necessitate context-specific interpretations and tailored solutions. Therefore, the workflow presented in this study should be viewed as a starting point for further exploration and refinement rather than a one-size-fits-all solution.

Moving forward, additional research and experimentation are needed to enhance the understanding of uncertainties in urban planning and develop robust, adaptive strategies capable of navigating future challenges. Planners can refine their approach and develop best practices for addressing uncertainty in urban contexts by conducting more case studies and experimenting with different methodologies and tools. This iterative process of learning and adaptation will be critical for building resilient and sustainable cities capable of thriving in an uncertain future.

4.4 General Discussion

Despite the limitations described, there is an urgent need for more applications of Decision Making Under Deep Uncertainty (DMDU) approaches and tools to assist professionals in urban and architectural fields. As discussed in Chap. 2, uncertainty about the future and the failure of mechanistic approaches do not restrict planners to merely guiding roles in transition. Planners and designers in the built environment should exert significant influence in transitioning toward a more integrated planning paradigm that combines disciplines, stakeholders, and communities while advocating for a regenerative and resilient built environment through active planning actions. Uncertainty about the future should not hinder the potential for action today to pursue a better future.

In this perspective, planners need to familiarize themselves with the concept of uncertainty and the different levels (see Chap. 2). In fact, not in all the situations planners and designers grope in the dark. The first two levels of uncertainty can usually still be managed with "conventional" planning (see Fig. 4.4), assessing which measure or strategy is the most robust for the specific context. The additional step is to understand what is necessary to monitor to have alternatives set in place. If the external changes are so severe that the system cannot cope with them (tipping point condition), planners should have alternatives set in place to foster adaptation and meet the objective still or seize opportunities that may arise from the dynamic context.

Tipping points have played a crucial role in informing adaptive planning across various disciplines [63, 76, 102–105]. Recognizing these critical junctures has

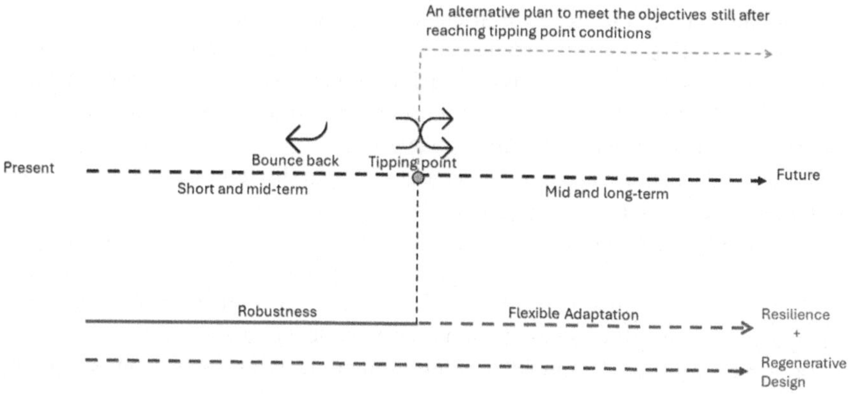

Fig. 4.4 This diagram conceptualizes the strategic continuum in urban planning from the present to future scenarios. It delineates a temporal progression toward resilience, starting with robust short- to mid-term planning strategies. A tipping point signifies a critical juncture where strategies may pivot to ensure adaptability. Post-tipping point, the diagram bifurcates to illustrate dual paths: immediate recovery tactics and alternative long-term plans that sustain the city's objectives under new conditions. Regenerative design and Resilience are not contrasting objectives but planning paradigms that can mutually benefit from their intersection

allowed professionals to identify when changes can lead to significant, sometimes irreversible, shifts in a system's state or behavior. This understanding is instrumental in developing flexible and robust strategies to accommodate sudden shifts and leverage them for innovation and transformation. Adaptive planning, thus, benefits from the awareness of tipping points, as it prepares systems not only to withstand or adapt to shocks but to use them as catalysts for positive change.

In the example provided in this chapter, exploratory scenarios were used to build narratives that described undesirable outcomes qualitatively. However, DMDU can couple qualitative and quantitative assessments into their methods to build adaptive plans that offer multiple routes toward unknown futures. This is essential not just for long-term planning but also for communicating the different actions that can be undertaken and the possible strengths and weaknesses that could be encountered.

Figure 4.5 is used as an illustrative example of the built environment; tapping into the information provided in Chap. 3, serves as a visual representation of adaptive planning strategies in flood-prone urban areas. It outlines initial planning solutions, such as dry-proof and wet-proof building technologies, under present conditions with shallow flooding and low population density. Anticipating future changes, it presents alternate pathways: Route B transitions to elevated and amphibious structures to accommodate deeper waters and increased population, while Route C employs nature-based solutions for water management without population growth. However, when environmental conditions deteriorate, and the population rises, Route D illustrates a shift to resilient building types combined with green infrastructure. Route E showcases a resilient strategy suitable for dense populations and severe flooding across time.

This schematic example hints that, with innovative planning paradigms and tools, urban planners and designers can create flexible, short-term designs with the capacity for future adaptation. Furthermore, it enhances communicative planning by evaluating diverse strategies under changing conditions and assessing if it is best to invest in the most robust strategy (E) with the risk of not encountering extreme situations in the future or if it is better to plan with mitigative strategies that have an end of validity (A) and be prepared for future changes.

Moreover, the detailed scenarios and methodologies discussed, exemplified by the illustrative timeline in Fig. 4.5, highlight the importance of visualizing and communicating complex planning concepts. Such visual tools are indispensable in engaging stakeholders, fostering a shared understanding of potential futures, and facilitating collaborative decision-making processes. They serve as a blueprint for immediate action and a roadmap for navigating long-term challenges and opportunities.

Furthermore, acknowledging that adaptive planning is not solely about mitigating risks but also about harnessing the potential for positive change is a compelling call to action for urban planners, policymakers, and communities. It invites a revaluation of conventional approaches and encourages the exploration of innovative solutions that prioritize resilience, regeneration, and the well-being of urban populations.

As the focus on flooding in this book demonstrates, the principles and practices of adaptive planning and DMDU are broadly applicable across various environmental and urban challenges. Whether it involves tackling water management,

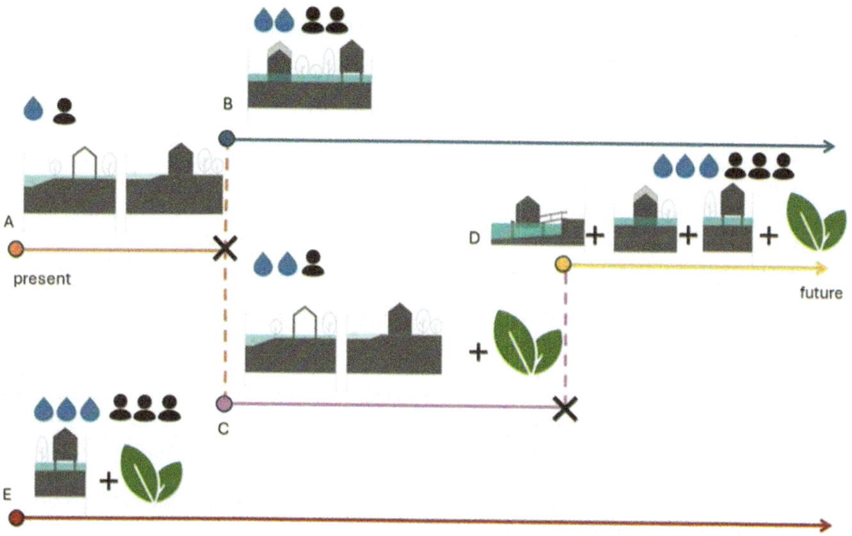

Fig. 4.5 This illustrative timeline depicts various urban planning scenarios from the present into the future. It visualizes the progression from conventional strategies (A) that may fail under certain conditions (X), toward more adaptive approaches (B, D) that anticipate and respond to changing circumstances, including potential tipping points. The figure highlights alternative pathways (C, E) incorporating resilience and regenerative design principles, demonstrating how planners can prepare for and adapt to unforeseen challenges while pursuing sustainable outcomes"

climate change, urban growth, or other complex challenges, integrating comprehensive and flexible planning strategies reflects the valuable role that foresight, collaboration, and innovation play in shaping resilient cities for the future.

In essence, the journey toward adaptive, resilient urban futures is both a challenge and an opportunity. It demands a reimagining of how we plan, design, and inhabit our urban spaces, guided by a deep understanding of the dynamics at play and a commitment to creating environments that thrive in the face of uncertainty and change.

References

1. M. Amer, T.U. Daim, A. Jetter, A review of scenario planning. Futures **46**, 23–40 (2013)
2. G. De Roo, Being or becoming? that is the question! confronting complexity with contemporary planning theory, in *A Planner's Encounter with Complexity*, ed. by G. de Roo, E.A. Silva (Ashgate, Farnham, 2010), pp. 19–40
3. V.A.W.J. Marchau, W.E.Walker, P.J.T.M. Bloemen, S.W. Popper, *Decision Making under Deep Uncertainty* (Springer, 2019), pp. 1–20
4. P. Schoemaker, Scenario planning: a tool for strategic thinking. Long Range Plann. **28**(3), 117 (1995)

5. T.J. Chermack, S.A. Lynham, W. Ruona, E.A., Scenario planning: a review of the literature. Fut. Res. Quart. **17**(2), 7–31 (2001)
6. H. Kahn, *Thinking About the Unthinkable* (Avon, 1964)
7. H. Kahn, A.J. Wiener, The next thirty-three years: a framework for speculation. Daedalus **96**(3), 705–732 (1967)
8. G. Ringland, *Scenario Planning: Mamaging the Future* (John Wiley & Sons, Chichester, 1998)
9. P. Carvalho, Scenarios: learning and acting from the future. Insight Foresight (2021)
10. F. Jahanishakib, S.H. Mirkarimi, A. Salmanmahiny, F. Poodat, Land use change modeling through scenario-based cellular automata Markov: improving spatial forecasting. Environ. Monit. Assess. **190**(6) (2018)
11. M. Godet, Introduction to la prospective. Futures **18**(2), 134–157 (1986)
12. G. Saliba, Windows for the mind: the use of scenario planning for managing uncertainty. J. Futuers Stud. **13**(February), 123–128 (2009)
13. C.A. Varum, C. Melo, Directions in scenario planning literature—a review of the past decades. Futures **42**(4), 355–369 (2010)
14. P.W.F. Van Notten, J. Rotmans, M.B.A. Van Asselt, D.S. Rothman, An updated scenario typology. Futures **35**, 423–443 (2003)
15. K. Cordova-Pozo, E.A.J.A. Rouwette, Types of scenario planning and their effectiveness: a review of reviews. Futures **149**(November), (2022)
16. S. Walton, P. O'Kane, D. Ruwhiu, Developing a theory of plausibility in scenario building: designing plausible scenarios. Futures **111**(March), 42–56 (2019)
17. L. Börjeson, M. Höjer, K.H. Dreborg, T. Ekvall, G. Finnveden, Scenario types and techniques: towards a user's guide. Futures **38**(7), 723–739 (2006)
18. G. Ducot, G.J. Lubben, A typology for scenarios. Futures **12**(1), 51–57 (1980)
19. U. Avin, R. Goodspeed, Using exploratory scenarios in planning practice: a spectrum of approaches. J. Am. Plan. Assoc. **86**(4), 403–416 (2020)
20. G. Lobosco, L. Tinti, B. Magagnoli, V. Mencarini, S. Mannucci, M. Ferrero, Landscape as a palimpsest for energy transition: correlations between the spatial development of energy-production infrastructure and climate-mitigation goals. Atmosphere **14**, 1–19 (2023)
21. J. Iverson Nassauer, R.C. Corry, Using normative scenarios in landscape ecology. Landsc. Ecol. **19**(4), 343–356 (2004)
22. A. Chakraborty, Enhancing the role of participatory scenario planning processes: lessons from reality check exercises. Futures **43**(4), 387–399 (2011)
23. G. Abou Jaoude, O. Mumm, V.M. Carlow, An overview of scenario approaches: a guide for urban design and planning. J. Plan. Lit. **37**(3), 467–487 (2022)
24. P. Allmendinger, M. Tewdwr-Jones, The communicative turn in urban planning: unravelling paradigmatic, imperialistic and moralistic dimensions. Space Polity **6**(1), 5–24 (2002)
25. J.E. Innes, D.E. Booher, *Planning with Complexity: An Introduction to Collaborative Rationality for Public Policy* (Routledge, London, 2010)
26. L.D. Hopkins, M.A. Zapata, *Engaging the Future: Forecasts, Scenarios, Plans, and Projects* (Lincoln Insitute of Land Policy, Cambridge, 2007)
27. R.E. Klosterman, Simple and complex models. Environ. Plan. B Plan. Des. **39**(1), 1–6 (2012)
28. R.E. Klosterman, Planning support systems: a new perspective on computer-aided planning. J. Plan. Educ. Res. **17**(1), 45–54 (1997)
29. U.P. Avin, J.L. Dembner, Getting scenario-building right. Planning **11**(November), 22–27 (2001)
30. U. Avin, R. Goodspeed, L. Murnen, From exploratory scenarios to plans: bridging the gap. Plan. Theory Pract. **23**(4), 637–646 (2022)
31. P. Schwartz, *The Art of the Long View. Planning for the Future in and Uncertain World* (Wiley, 1997)
32. D. Myers, A. Kitsuse, Constructing the future in planning: a survey of theories and tools. J. Plan. Educ. Res. **19**(3), 221–231 (2000)

33. J. Kwakkel, M. Haasnoot, W. Walker, Computer assisted dynamic adaptive policy design for sustainable water management in river deltas in a changing environment. International Congress on Environmental Modelling and Software Managing Resources of a Limited Planet, Sixth Biennial Meeting, Leipzig, Germany (2012), pp. 1801–1810
34. R. Bradfield, G. Wright, G. Burt, G. Cairns, K. Van Der Heijden, The origins and evolution of scenario techniques in long range business planning. Futures **37**(8), 795–812 (2005)
35. P. Bishop, A. Hines, T. Collins, The current state of scenario development: an overview of techniques. Foresight **9**(1), 5–25 (2007)
36. J. Marlow, H. Oliver, R. Quay, R. Marra, Integrating exploratory scenario planning into a municipal general plan update. Lincoln Institute of Land Policy (2015)
37. G.A. Riddell, H. van Delden, H.R. Maier, A.C. Zecchin, Exploratory scenario analysis for disaster risk reduction: considering alternative pathways in disaster risk assessment. Int. J. Disast. Risk Reduct. **39**, 101230 (2019)
38. A. Chakraborty, N. Kaza, G.J. Knaap, B. Deal, Robust plans and contingent plans. J. Am. Plan. Assoc. **77**(3), 251–266 (2011)
39. M. Haasnoot, H. Middelkoop, A. Offermans, E. Van Beek, W.P.A. van Deursen, Exploring pathways for sustainable water management in river deltas in a changing environment. Clim. Change **115**(3–4), 795–819 (2012)
40. A. Chakraborty, A. McMillan, Scenario planning for urban planners: toward a practitioner's guide. J. Am. Plan. Assoc. **81**(1), 18–29 (2015)
41. L. Schueller, L. Booth, K. Fleming, J. Abad, Using serious gaming to explore how uncertainty affects stakeholder decision-making across the science-policy divide during disasters. Int. J. Disast. Risk Reduct. **51**(June), 101802 (2020)
42. W.E. Walker et al., Defining uncertainty: a conceptual basis for uncertainty management in model-based decision support. Integr. Assess. **4**(1), 5–17 (2003)
43. J.H. Kwakkel, W.E. Walker, V.A.W.J. Marchau, Classifying and communicating uncertainties in model-based policy analysis. Int. J. Technol. Policy Manag. **10**(4), 299–315 (2010)
44. R.J. Lempert, S.W. Popper, S.C. Bankes, *Shaping the Next One Hundred Years: New Methods for Quantitative, Long-Term Policy Analysis* (RAND Corporation, Santa Monica, CA, 2003)
45. W.E. Walker, S.A. Rahman, J. Cave, Adaptive policies, policy analysis, and policy-making. Eur. J. Oper. Res. **128**(2), 282–289 (2001)
46. W.E. Walker, V.A.W.J. Marchau, D. Swanson, Addressing deep uncertainty using adaptive policies: introduction to section 2. Technol. Forecast. Soc. Change **77**(6), 917–923 (2010)
47. J.H. Kwakkel, M. Haasnoot, W.E. Walker, Developing dynamic adaptive policy pathways: a computer-assisted approach for developing adaptive strategies for a deeply uncertain world. Clim. Change. **132**(3), 373–386 (2015)
48. M. Haasnoot, J.H. Kwakkel, W.E. Walker, J. ter Maat, Dynamic adaptive policy pathways: a method for crafting robust decisions for a deeply uncertain world. Glob. Environ. Chang. **23**(2), 485–498 (2013)
49. J.H. Kwakkel, M. Haasnoot, W.E. Walker, Comparing robust decision-making and dynamic adaptive policy pathways for model-based decision support under deep uncertainty. Environ Model Softw. **86**, 168–183 (2016)
50. R.J. Lempert, D.G. Groves, S.W. Popper, S.C. Bankes, A general, analytic method for generating robust strategies and narrative scenarios. Manag. Sci. **52**(4), 514–528 (2006)
51. J.H. Kwakkel, M. Haasnoot, Supporting DMDU: A taxonomy of approaches and tools. in *Decision Making Under Deep Uncertainty: From Theory to Practice*, vol. A, ed. by W.J. Marchau, W.E. Walker, P.J.T.M. Bloemen, S.W. Popper (Springer International Publishing, Cham, 2019), pp. 355–374
52. P. Goodwin, G. Wright, The limits of forecasting methods in anticipating rare events. Crit. Stud. Secur. **2**(2), 210–222 (2010)
53. V.A.W.J. Marchau, W.E. Walker, P.J.T. Bloemen, S. Popper, *Decision Making Under Deep Uncertainty* (Springer International Publishing, 2019).
54. E. Diehl, J.D. Sterman, Effects of feedback complexity on dynamic decision making. Organ. Behav. Hum. Decis. Process. **62**(2), 198–215 (1995)

55. J.D. Sterman, All models are wrong: reflections on becoming a systems scientist. Syst. Dyn. Rev. **18**(4), 501–531 (2002)
56. S. Bankes, Models as lab equipment: science from computational experiments. Comput. Math. Organ. Theory **15**(1), 8–10 (2009)
57. J.H. Kwakkel, E. Pruyt, Exploratory modeling and analysis, an approach for model-based foresight under deep uncertainty. Technol. Forecast. Soc. Change **80**(3), 419–431 (2013)
58. G. Holtz et al., Prospects of modelling societal transitions: position paper of an emerging community. Environ. Innov. Soc. Transit. **17**, 41–58 (2015)
59. R.J. Lempert et al., *Ensuring Robust Flood Risk Management in Ho Chi Minh City*. World Bank (2013), pp. 1–63
60. V.A.W.J. Marchau, W.E. Walker, P.J.T.M. Bloemen, S.W. Popper, Introduction. in *Decision Making under Deep Uncertainty: From Theory to Practice*, vol. A, ed. by W.J. Marchau, W.E. Walker, P.J.T.M. Bloemen, S.W. Popper (Springer International Publishing, Cham, 2019), pp. 1–20
61. S. Bankes, W.E. Walker, J.H. Kwakkel, Exploratory modeling and analysis. Encycl. Oper. Res. Manag. Sci. **2**(1), 532–537 (2013)
62. M.C.B. Stanton, K. Roelich, Decision making under deep uncertainties: a review of the applicability of methods in practice. Technol. Forecast. Soc. Change **171**(June), 120939 (2021)
63. W.E. Walker, M. Haasnoot, J.H. Kwakkel, Adapt or perish: a review of planning approaches for adaptation under deep uncertainty. Sustainability. **5**(3), 955–979 (2013)
64. S. Bankes, Exploratory modeling for policy analysis. Oper. Res. **41**(3), 435–449 (1993)
65. M. Jaxa-Rozen, J. Kwakkel, Tree-based ensemble methods for sensitivity analysis of environmental models: a performance comparison with Sobol and Morris techniques. Environ Model Softw. **107**, 245–266 (2018)
66. B.P. Bryant, R.J. Lempert, Thinking inside the box: a participatory, computer-assisted approach to scenario discovery. Technol. Forecast. Soc. Change **77**(1), 34–49 (2010)
67. J.H. Kwakkel, A generalized many-objective optimization approach for scenario discovery. Futures Foresight Sci. **1**(2), e8 (2019)
68. C. McPhail, H.R. Maier, J.H. Kwakkel, M. Giuliani, A. Castelletti, S. Westra, Robustness metrics: how are they calculated, when should they be used and why do they give different results? Earth's Future **6**(2), 169–191 (2018)
69. L.J. Savage, The theory of statistical decision. J. Am. Stat. Assoc. **46**(253), 55–67 (1951)
70. H.A. Simon, Rational choice and the structure of the environment. Psychol. Rev. **63**(2) (1956)
71. M.C.B. Stanton, K. Roelich, Decision making under deep uncertainties: a review of the applicability of methods in practice. Technol. Forecast. Soc. Change. **171**(April), 120939 (2021)
72. C. Hamarat, J.H. Kwakkel, E. Pruyt, Adaptive robust design under deep uncertainty. Technol. Forecast. Soc. Change **80**(3), 408–418 (2013)
73. E.A. Eriksson, K.M. Weber, Adaptive foresight: navigating the complex landscape of policy strategies. Technol. Forecast. Soc. Change **75**(4), 462–482 (2008)
74. M.K. Webber, C. Samaras, A review of decision making under deep uncertainty applications using green infrastructure for flood management. Earth's Future **10**(7) (2022)
75. D. Groves, E. Bloom, J.R. Fischbach, D. Knopman, Adapting to a changing colorado river: making future water deliveries more reliable through robust management strategies, in *AGU Fall Meeting Abstracts*, vol. 2013 (2013), pp. H41K-1407
76. M. Haasnoot, A. Warren, J.H. Kwakkel, Dynamic adaptive policy pathways (DAPP). in *Decision Making under Deep Uncertainty: From Theory to Practice*, vol. A, ed. by W.J. Marchau, W.E. Walker, P.J.T.M. Bloemen, S.W. Popper (Springer International Publishing, Cham, 2019), pp. 71–92
77. J. Lawrence, M. Haasnoot, L. McKim, D. Atapattu, A. Stroombergen, G. Campbell, Dynamic adaptive policy pathways (DAPP): from theory to practice. in *Decision Making Under Deep Uncertainty: From Theory to Practice*, ed. by V.A.W.J. Marchau, W.E. Walker, P.J.T.M. Bloemen, S.W. Popper (Springer International Publishing, Cham, 2019), pp. 187–199

78. J.W.G.M. Van der Pas, W.E. Walker, V.A.W.J. Marchau, B. van Wee, J.H. Kwakkel, Operationalizing adaptive policymaking. Futures **52**, 12–26 (2013)
79. S.E. Werners, R.M. Wise, J.R.A. Butler, E. Totin, K. Vincent, Adaptation pathways: a review of approaches and a learning framework. Environ. Sci. Policy **116**(January), 266–275 (2021)
80. S. Mannucci, J.H. Kwakkel, M. Morganti, M. Ferrero, Exploring potential futures: evaluating the influence of deep uncertainties in urban planning through scenario planning: a case study in Rome, Italy. Futures **154**(September), 103265 (2023)
81. J.H. Kwakkel, M. Jaxa-Rozen, Improving scenario discovery for handling heterogeneous uncertainties and multinomial classified outcomes. Environ. Model Softw. **79**, 311–321 (2016)
82. F. Recanatesi, A. Petroselli, Land cover change and flood risk in a peri-urban environment of the metropolitan area of Rome (Italy). Water Resour. Manag. **34**, 4399–4413 (2020)
83. E. Aureli Cutillo, F. Mingella Calvosa, *Abitare a Roma. Urbanizzazione e crescita urbana*, Studi e ri. Franco Angeli Editore (1989)
84. P. Berdini, *Breve storia dell'abuso edilizio in Italia.* (Roma, 2010)
85. F. Chiodelli et al., The production of informal space: a critical atlas of housing informalities in Italy between public institutions and political strategies. Prog. Plann. **149**, 100495 (2021)
86. I. Clough Marinaro, M. Solimene, Navigating the (in)formal city: Roma, urban life and governance in Rome. Cities **96**, 102402 (2020)
87. C. Cellamare, Politiche e processi dell'abitare nella città abusiva/informale romana, Arch. di Stud. Urbani e Reg. Fr. Angeli Ed., no. Fascicolo 97–98/2010, 145–167 (2010)
88. R. BV, M Etronamica (2012)
89. J. Van Vliet, A.K. Bregt, D.G. Brown, H. Van Delden, A review of current calibration and validation practices in land-change modeling environmental modelling & software a review of current calibration and validation practices in land-change modeling. Environ. Model Softw. **82**(May), 174–182 (2016)
90. J. Van Vliet, A.K. Bregt, A. Hagen-zanker, Revisiting kappa to account for change in the accuracy assessment of land-use change models. Ecol. Model. **222**(8), 1367–1375 (2011)
91. A. Hagen, Fuzzy set approach to assessing similarity of categorical maps. Int. J. Geogr. Inform. Sci. **8816** (2010)
92. J.H. Kwakkel, The exploratory modeling workbench: an open source toolkit for exploratory modeling, scenario discovery, and (multi-objective) robust decision making. Environ Model. Softw. **96**, 239–250 (2017)
93. J. C. Helton, F. J. Davis, Latin Hypercube Sampling and the propagation of uncertainty in analyses of complex systems. Reliab. Eng. Syst. Saf. **81**(1), 23-69 (2002)
94. C. Helgeson, Structuring decisions under deep uncertainty. Topoi **39**(2), 257–269 (2020)
95. J. Rozenberg, C. Guivarch, R. Lempert, S. Hallegatte, Building SSPs for climate policy analysis: a scenario elicitation methodology to map the space of possible future challenges to mitigation and adaptation. Clim. Change. **122**(3), 509–522 (2014)
96. M.D. Gerst, P. Wang, M.E. Borsuk, Discovering plausible energy and economic futures under global change using multidimensional scenario discovery. Environ. Model. Softw. **44**, 76–86 (2013)
97. E.A. Moallemi, J. Kwakkel, F.J. de Haan, B.A. Bryan, Exploratory modeling for analyzing coupled human-natural systems under uncertainty, Glob. Environ. Chang. **65**(September), 102186 (2020)
98. M.R. Ackermann, J. Blömer, D. Kuntze, C. Sohler, Analysis of agglomerative clustering. Algorithmica **69**(1), 184–215 (2014)
99. S. Theodoridis K. Koutroumbas, Clustering algorithms II: hierarchical algorithms. in *Pattern Recognition* (Academic Press, 2009), pp. 653–700
100. X. Jin, J. Han, K-Medoids clustering. in *Encyclopedia of Machine Learning*, ed. By C. Sammut, G. I. Webb (Springer US, Boston, MA, 2010), pp. 564–565
101. L. Borges, M. Hazin, A. Teixeira, D. Almeida, Multidimensional flood risk management under climate changes: bibliometric analysis, trends and strategic guidelines for decision-making in urban dynamics. Int. J. Disast. Risk Reduct. **50**(June), 101865 (2020)

102. J.C.J. Kwadijk et al., Using adaptation tipping points to prepare for climate change and sea level rise: a case study in the Netherlands. Wiley Interdiscip. Rev. Clim. Change **1**(5), 729–740 (2010)
103. S.E. Werners, S. Pfenninger, E. van Slobbe, M. Haasnoot, J.H. Kwakkel, R.J. Swart, Thresholds, tipping and turning points for sustainability under climate change. Curr. Opin. Environ. Sustain. **5**(3–4), 334–340 (2013)
104. J.W. Hall, H. Harvey, L.J. Manning, Adaptation thresholds and pathways for tidal flood risk management in London. Clim. Risk Manag. **24**(April), 42–58 (2019)
105. N. Rocle et al., Paving the way to coastal adaptation pathways: an interdisciplinary approach based on territorial archetypes. Environ. Sci. Policy **110**, 34–45 (2020)

Chapter 5
Conclusions and Perspectives

Abstract The pressing need for an evolved planning paradigm in architectural and urban fields is emphasized in response to climate change and the complexities of the built environment. This transition involves moving from traditional, deterministic planning models to dynamic, adaptive strategies that prioritize resilience and sustainability in urban development. Planners are repositioned as facilitators of urban transformation, engaging with multifaceted processes and diverse stakeholders to foster adaptive and resilient urban environments.The integration of multiscale and multidisciplinary approaches is essential to effectively address climate adaptation challenges, particularly urban flooding. A critical examination reveals a gap between theoretical frameworks and practical implementation, advocating for enhanced tools, methodologies, and resources to adequately equip planners. A comprehensive approach is called for, one that embraces flexibility, continuous evaluation, and adaptive strategies to ensure the long-term resilience and sustainability of urban environments. By integrating advanced planning tools and methodologies, urban planners can better anticipate and respond to future challenges, ensuring cities can thrive in the face of uncertainty.

Keywords Climate change · Urban resilience · Built environment · Adaptive strategies · Multi-disciplinary approaches · Adaptive planning

5.1 Conclusions

Climate change and the inherent complexity of the built environment are two key factors that underscore the necessity of evolving the planning paradigm in architectural and urban fields. This evolution is characterized by a significant shift from traditional models toward dynamic, innovative strategies that address the multifaceted challenges climate change presents to urban environments. This transformation significantly affects the planning profession, transitioning from a deterministic, mechanical viewpoint where planners shape the world through visionary perspectives to a more adaptive and responsive position. In this evolved role, they serve as

S. Mannucci, *Climate Adaptation in Urban Planning*,
SpringerBriefs in Architectural Design and Technology,
https://doi.org/10.1007/978-981-97-4106-9_5

facilitators of urban transformation and adaptation. This underscores the importance of recognizing the complex, interdependent processes that underlie how urban environments evolve, leading to non-linear development trajectories. In such a dynamic context, the adaptability of urban areas to both anticipated and unforeseen changes becomes crucial for maintaining their vitality.

Consequently, planners are encouraged to shift their focus from the content and process of planning to fostering conditions conducive to development. By engaging with emerging trends, linking projects and local coalitions, and promoting innovation, planners co-create the city with a range of actors, navigating the intricacies of adaptive urban transformation. This paradigm shift invites a reevaluation of planning strategies to accommodate the fluidity of urban development, ensuring cities remain resilient and adaptive in the face of change [1].

The emerging narratives, delving into the current planning paradigms, have woven together the importance of an integrated approach that melds sustainable, regenerative, resilient, and flexible planning principles. Such a holistic strategy is indispensable for tackling the complex and dynamic interactions between urban systems and climate change. It calls for the adoption of multi-disciplinary and multi-scale strategies to evaluate the most robust short-term planning strategies, leaving room for future long-term adaptation without undermining the original objectives of a plan. However, it is clear that the planners are not sufficiently equipped to put into practice the theoretical frameworks and innovative strategies required to navigate these evolving challenges. Planners often find themselves at the forefront of the battle against climate change. Nevertheless, the tools, methodologies, and resources at their disposal are not always suitable to assist them, lacking the specificity and flexibility needed to address the unique complexities of modern urban environments.

Specifically, this book addresses one of the most compelling phenomena related to climate change, namely, urban flooding. Chapter 3 presents the multi-scale approach that the literature promotes as best practices to implement resilience and adaptation, illustrating the main approaches at the different scales, varying from normative or policy-based interventions at a wider scale and the technical solutions at smaller scales where planners can intervene with context-specific design measures. However, despite the best efforts, a critical gap emerges as the different scales do not communicate as effectively as they should. Integrating a multi-scale and multi-disciplinary application to pursue climate adaptation in urbanized flood-prone contexts still faces several challenges. The comprehensive exploration of climate change adaptation in the built environment, as delineated, reveals the burgeoning emphasis on resilience and adaptive planning amidst growing research. Yet, critical limitations and gaps within this extensive body of work call for urgent attention. The multi-scale strategy for flood management, pivotal in climate change adaptation, faces challenges that underscore the complexity of forecasting and managing flood risks, technical hurdles in multi-scale modeling, and socioeconomic influences on strategy implementation. This discourse highlights the necessity for a local perspective in strategy evaluation, the ambiguity surrounding adaptive timing and methods for the built environment, and the predominance of studies focusing on narrow adaptation facets. An interdisciplinary approach, merging insights from architecture, engineering, environmental science, and social sciences, emerges as essential, advocating for integrated tools and methodologies to foster holistic, long-term adaptation planning. This conclusion underscores the importance of economic feasibility, resilience, environmental sustainability, and regeneration in planning, emphasizing the need for continuous evaluation and adaptability in the planning framework to navigate the uncertainties of long-term environmental and societal changes.

As discussed in Chaps. 2 and 4, scenario planning has been proven to be a valuable aid in several disciplines that require planning for the future. In light of the growing uncertainty surrounding the future of urban areas, planners are now challenged with the crucial task of making well-informed decisions in the present day. While exploratory scenarios have proven to be of significant value for forward-thinking planners, their application remains predominantly seen in sectors such as land use and infrastructure planning [2, 3], water management [4], and energy transition [5]. These applications, though crucial, illustrate the broader potential for exploratory scenarios across various aspects of urban development.

Currently, exploratory scenarios are being implemented on a larger scale, and despite the recognition that climate-related issues, such as flooding, must be tackled with a multi-scale, multi-disciplinary approach (see Chap. 3), each scale and discipline has its own set of tools and planning perspectives that are hardly communicating. This nature of problem-solving poses a significant barrier to effective communication and integration of efforts across different sectors.

In order to overcome these barriers, there is a pressing need for a more cohesive framework that facilitates cross-disciplinary dialogue and sharing of insights. Urban planners can leverage a broader array of strategies to address complex climate-related challenges by fostering an environment where various planning tools and perspectives are harmonized. This integrated approach enhances the effectiveness of urban planning initiatives and ensures a more resilient and sustainable urban future. Bridging the communication gap between different scales and disciplines is essential for successfully implementing exploratory scenarios in urban planning. Through collaborative efforts and the establishment of common ground, urban planners can more effectively utilize exploratory scenarios to anticipate and prepare for a wide range of future challenges, ensuring cities are better equipped to adapt and flourish in the face of uncertainty.

Drawing upon the intricate exploration of scenario planning and its application within the realm of urban planning, Chap. 4 has highlighted the critical role of scenario planning as a strategic tool for navigating the complexities and uncertainties that characterize the built environment, particularly in the context of climate change.

With its forward-looking and dynamic approach, scenario planning provides planners, with a structured framework to envision alternative futures, thereby offering a valuable perspective in the strategic decision-making process. This approach enables the consideration of a wide array of possible scenarios, facilitating a more robust and adaptable planning process better suited to respond to the unpredictable challenges of our times.

The exploration of Decision Making Under Deep Uncertainty (DMDU) within this context further underscores the necessity for innovative methodologies and tools to enhance urban spaces' resilience and adaptability. By integrating scenario planning into the decision-making processes, urban planners are equipped with the means to foresee and prepare for various future conditions, ensuring that urban environments are sustainable and capable of thriving amid the uncertainties brought forth by climate change.

5.2 Future Outlooks

In light of the constantly evolving and intricately intertwined challenges surrounding the future of urban development, urban planners face the daunting task of navigating a landscape characterized by unprecedented complexity. This reality underscores the urgent necessity for planners to embrace a diverse range of tools and methodologies capable of effectively addressing the multifaceted nature of urban development. Among these tools, parametric, computational, and algorithmic design methodologies emerge as promising approaches, showing potential with the integration of DMDU tools and methods.

Urban parametric design [6] represents a transformative approach within urban planning and design, integrating computational methods to tackle intricate urban challenges. This methodology harnesses the capabilities of parametric modeling to foster collaborative and participatory processes, thereby accommodating diverse stakeholders [7]. A pivotal aspect of this approach is its adoption of a multi-scale strategy, which delves into the nuances of urban microclimates and energy consumption, aiming to foster sustainable urban environments [8, 9].

The potential of parametric design to revolutionize planning through heightened adaptability, sustainability, and efficiency is undeniable. As planning paradigms continue to evolve in the face of complexity and environmental changes, parametric design emerges as a forward-thinking methodology that amalgamates the precision of computational analysis with human-centric design, propelling strides toward resilient and regenerative urban futures. The scientific literature in the field mainly focuses on reducing heat stress or energic district optimization by integrating parametric design with multi-objective optimization. However, computational tools and platforms can integrate the analyses and assessment of several environmental issues and/or optimize solutions considering a fixed number of contrasting objectives [10].

In this perspective, given the built environment's complexity, an interesting future direction would be testing many-objective optimization in addressing complex engineering design challenges characterized by numerous objectives [11]. This methodology enriches the decision-making process by furnishing a comprehensive analysis of trade-offs between conflicting goals, thereby facilitating the identification of optimal design solutions. Many-objective optimization encounters the challenge of diminishing returns in performance improvement as the number of objectives escalates, a limitation less pronounced in multi-objective optimization. This underscores the necessity for advanced computational techniques and visualization tools to comprehensively explore and comprehend the solution space, underscoring the demand for specialized approaches in navigating the intricacies associated with many-objective optimization scenarios.

Furthermore, students in the field should have access to education and training programs to cover the breadth of knowledge needed to navigate the complexity of providing short-term planning strategies that can and should be part of a broader plan that acknowledges that conditions may change, requiring further intervention from planners in the future.

It also involves the development of new, dynamic planning tools that can handle the complexity of cities and neighborhoods and the uncertainty of future scenarios. Such tools should enable planners to simulate various outcomes, assess the viability of different design and policy options, and make informed decisions prioritizing sustainability, equity, and resilience.

In conclusion, the intricate and ever-evolving challenges inherent in urban planning and design require a comprehensive and forward-thinking approach from planners. The unprecedented complexity demands the adoption of a diverse range of tools and methodologies capable of addressing the multifaceted nature of urban development. The challenges that climate change poses require the development of dynamic planning tools capable of handling the intricacies of cities and neighborhoods alongside the uncertainty of future scenarios. These tools and methodologies should empower planners to simulate various outcomes, evaluate the feasibility of diverse design and policy options, and make informed decisions prioritizing sustainability, regeneration, equity, and resilience.

References

1. W. Rauws, Embracing uncertainty without abandoning planning: exploring an adaptive planning approach for guiding urban transformations. disP Plan. Rev. **53**(1), 32–45 (2017)
2. A. Ariza-Álvarez, J.A. Soria-lara, F. Aguilera-benavente, Re-thinking the role of exploratory scenarios for adaptive policymaking: an application for land use and transport planning. Futures **149**, 1–13 (2023)
3. O. Roman, A.B. Elvarsson, B.T. Adey, Exploratory modelling for transport infrastructure planning under future uncertainty. Life-Cycle Struct. Infrastruct. Syst. 2580–2587 (2023)
4. C. Urich, W. Rauch, Exploring critical pathways for urban water management to identify robust strategies under deep uncertainties. Water Res. **66**, 374–389 (2014)
5. E.A. Moallemi, F. de Haan, J. Kwakkel, L. Aye, Narrative-informed exploratory analysis of energy transition pathways: a case study of India's electricity sector. Energy Policy **110**(February), 271–287 (2017)
6. J. Gil, J.P. Duarte, Towards an urban design evaluation framework integrating spatial analysis techniques in the parametric urban design process, Proc. Int. Conf. Educ. Res. Comput. Aided Archit. Des. Eur. 257–264 (2008)
7. N. Steinø, M.B. Yıldırım, M. Özkar, Parametric design strategies for collaborative and participatory urban design. Proc. Int. Conf. Educ. Res. Comput. Aided Archit. Des. Eur. **1**, 195–203 (2013)
8. C. Schneider, A. Koltsova, G. Schmitt, W.P. Strasse, Components for parametric urban design in grasshopper. From street network to building geometry. Building 68–75 (2010)

9. N. Abdollahzadeh, N. Biloria, Urban microclimate and energy consumption: a multi-objective parametric urban design approach for dense subtropical cities. Front. Archit. Res. **11**(3), 453–465 (2022)
10. H. Kim, M.J. Clayton, A multi-objective optimization approach for climate-adaptive building envelope design using parametric behavior maps. Build. Environ. **185**, 107292 (2020)
11. P.J. Fleming, R.C. Purshouse, R.J. Lygoe, Many-objective optimization: an engineering design perspective. Lect. Notes Comput. Sci. **3410**(January), 14–32 (2005)

SPRINGER NATURE

GPSR Compliance

The European Union's (EU) General Product Safety Regulation (GPSR) is a set of rules that requires consumer products to be safe and our obligations to ensure this.

If you have any concerns about our products, you can contact us on ProductSafety@springernature.com

In case Publisher is established outside the EU, the EU authorized representative is:

Springer Nature Customer Service Center GmbH
Europaplatz 3
69115 Heidelberg, Germany

The manufacturer's authorised representative in the EU is Springer Nature Customer Service Centre GmbH, Europaplatz 3, 69115 Heidelberg, Germany. If you have any concerns regarding our products, please contact ProductSafety@springernature.com

Printed and bound by CPI Group (UK) Ltd, Croydon, CR0 4YY
01/10/2026
02225880-0007